PRODUCING
HIGH-QUALITY FIGURES

USING SAS/ GRAPH® *and* ODS GRAPHICS PROCEDURES

PRODUCING HIGH-QUALITY FIGURES
USING SAS/ GRAPH® *and* ODS GRAPHICS PROCEDURES

CHARLIE CHUNHUA LIU

Kythera Biopharmaceuticals

Westlake Village, CA, USA

CRC Press
Taylor & Francis Group
Boca Raton   London   New York

CRC Press is an imprint of the
Taylor & Francis Group, an **informa** business

A CHAPMAN & HALL BOOK

CRC Press
Taylor & Francis Group
6000 Broken Sound Parkway NW, Suite 300
Boca Raton, FL 33487-2742

© 2015 by Taylor & Francis Group, LLC
CRC Press is an imprint of Taylor & Francis Group, an Informa business

No claim to original U.S. Government works

Printed on acid-free paper
Version Date: 20141229

International Standard Book Number-13: 978-1-4822-0701-9 (Paperback)

Visit the Taylor & Francis Web site at
http://www.taylorandfrancis.com

and the CRC Press Web site at
http://www.crcpress.com

Contents

Preface

Purpose of the Book

Most statisticians and SAS programmers, including some experienced SAS users, might consider SAS a poor tool for producing high-quality figures. This is a misconception and not true. SAS can produce high-quality figures that meet publication requirements if the right graph formats, options, and fonts are used.

This book illustrates the principles for producing high-quality figures and demonstrates how to apply those principles to produce various types of commonly used figures and plots with practical examples using SAS/GRAPH and ODS Graphics procedures. (SAS/GRAPH is the graphics and presentation component of the SAS software suite, and the ODS Graphics procedures, sometimes called ODS Statistical Graphics procedures, use output delivery system graphics functionality to produce plots. It is part of Base SAS in version 9.3.) It will be a useful and practical guide for statisticians and SAS programmers who want to present research data in high-quality figures to meet publication requirements in academic institutions and various industries, including pharmaceuticals, agriculture, financial institutions, and so on.

This book focuses on the details of choosing the right figure formats and fonts to produce high-quality figures to meet publication requirements for inclusion in a clinical study report (CSR) for a pharmaceutical company for manuscript submissions in general scientific communities.

The book provides detailed instructions and SAS programs using procedures in both SAS/GRAPH and ODS Graphics to produce practical sample figures in listing graphics formats (PS, EPS, EMF, etc.) and in ODS document files (PDF, RTF). Readers can easily modify the SAS programs included in the book to produce high-quality figures to meet their own needs.

ODS Graphics procedures (SGPLOT, SGSCATTER, etc.), which were previously part of SAS/GRAPH in SAS 9.2, are known as statistical graphics (SG) procedures, and have been moved to base SAS in SAS 9.3. While the ODS Graphics procedures provide a more convenient and efficient way to produce certain types of figures, especially multicell and multipanel plots, the traditional procedures of SAS/GRAPH are still very versatile and might be preferred to produce highly customized, stand-alone figures because they provide significant flexibility to control the appearance of the output. Most figures in this book are produced using procedures in both SAS/GRAPH and ODS Graphics and the features of both systems, and the pros and cons are compared and discussed. The SAS programs for the sample figures were produced in SAS 9.3.

It is impossible for a single book to cover every procedure in SAS/GRAPH and ODS Graphics. The procedures used in the examples in the book include the following

SAS/GRAPH procedures:

- PROC GPLOT
- PROC GCHART
- PROC GREPLAY

ODS Graphics procedures:

- PROC SGPLOT
- PROC SGPANEL
- PROC SGSCATTER

PROC UNIVARIATE in Base SAS is used to produce histograms.

Book Structure and Suggested Reading Order

The first two chapters discuss the kinds of high-quality figures that are required for publication, the principles for producing those high-quality figures, a comparison of figures in vector and bitmap format, and how to choose the right formats and fonts. These two chapters are suggested for all readers.

Chapters 3 to 13 discuss and demonstrate how to produce some of the most commonly used types of plots, including line plots, scatter and jittered scatter plots, line-up jittered scatter plots, thunderstorm and raindrop scatter plots, spaghetti plots, bar charts, box plots, forest plots, survival plots, waterfall plots, histograms, and Bland–Altan plots for agreement analyses, and so on. The line-up jittered scatter plots and the thunderstorm/raindrop scatter plots are new plot types that have not been published before. Readers can go directly to individual chapters to learn how to produce the specific types of figures they are interested in generating. Practical examples and SAS programs are included in each chapter.

Chapter 14 illustrates how to produce some sample figures in SAS ODS Graphic designer, a convenient tool for producing figures without having to write SAS programs. This might be useful for readers who are not very familiar with SAS programming and SAS Graphics.

Chapter 15 discusses how to include all of the SAS programs that produced the sample figures in this book in an Enterprise Guide (EG) project. Some advantages and features in using EG to produce figures are discussed. A point-and-click feature to produce simple figures in EG is also illustrated.

Disclaimer on Sample Figures

All the examples in the clinical research area in the book are based on actual clinical scenarios, mostly in the glaucoma research area when author was a project statistician at Allergan, Inc. However the data used for the sample figures are simulated. Therefore, please do not make any clinical interpretations based on them. Figures in nonclinical areas—agriculture, finance, and so on—are based on published data.

Acknowledgments

The author would like to acknowledge the following people and institutions for their invaluable review comments and excellent editing suggestions.

Mark Chang, PhD, vice president at AMAG Pharmaceuticals and adjunct professor at Boston University for his review and valuable comments on the book proposal. Sunil Gupta, best-selling SAS author and global corporate trainer, for his review of the book manuscripts and valuable comments on the layout and code in each chapter. Tracy Turschman, senior director of statistical programming at Vertex Pharmaceuticals, for his time and valuable review comments on the book proposal and the book manuscripts. Lily Xu, Associate Director of Statistical Programming at Shire Pharmaceuticals, for her testing of the SAS programs and valuable comments and suggestions on the SAS codes.

SAS Institute, for review and feedback on the manuscripts and for providing free SAS licenses for the book project, and to North Carolina State University where the author learned SAS programming and received his degree in statistics.

The author would also like to thank Richard Cooper, PhD, professor at North Carolina State University, for his encouragement and especially for his lifelong friendship and support. Dr. Cooper is the author's graduate student adviser in the PhD program and has provided invaluable support in the author's academic development and life.

Last but not least, the author would like to thank his family for their understanding and support during the writing and finalization of the book, mostly in the long evenings and weekends.

Author Biography

Charlie Chunhua Liu, PhD (劉春華博士), was born in the village of Shihu, Auyuan County, Jiangxi Province, China (中國江西省安遠縣蔡坊鄉仕湖上村). He received a PhD in crop science (turfgrass management) and a master's in statistics (biomedical concentration) from North Carolina State University.

Dr. Liu has worked as an SAS programmer and project statistician in several research institutions and pharmaceutical companies, including the US Environmental Protection Agency (EPA), the National Institute of Statistical Sciences (NISS), the Washington University Medical School at St. Louis, Eli Lilly and Company, and Allergan Inc. He is now an Associate Director of Biostatistics at Kythera Biopharmaceutucals Inc., Westlake Village, CA.

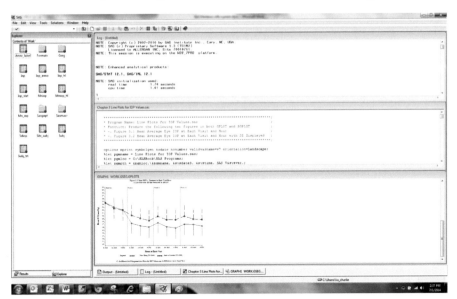

An SAS Programmer's Prayer
> Oh, Lord, please listen to an SAS programmer's prayer
> May SAS smile upon me
> May its EDITOR always show me the right codes
> May its OUTPUT produce the tables and listings that I want
> May its LOG never give me anything in red
> May its GRAPH generate figures beautiful and meaningful
> Oh, and most of all,
> **May I enjoy life as much as I enjoy coding in SAS.**
Amen!

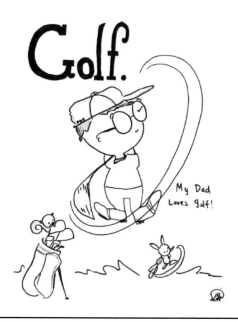

數字人生
　　讀好書交諍友
　　喝點酒打高球
　　一二好友足矣
　　三四兩酒不多
　　五六本書嫌少
　　七八拾桿高手
　　九十多歲老翁
　　百味人生苦樂
　　千種生活歷煉
　　萬般感恩歸主

A Life of Numbers
　　Read good books.
　　Make true friends.
　　Drink nice wine.
　　Play wonderful golf.
　　Making *One* or *Two* true friends is good enough.
　　Drinking *Three* or *Four* ounces of liquor could just be fine.
　　Reading *Five* or *Six* books in a year is really not good enough.
　　Playing golf with *Seventy* to *Eighty* strokes in a round is my goal.
　　Live into my *Nineties*, with *Hundreds* of life experiences
　　　　and *Thousands* of life pleasures.
Finally, rest in my Lord with *Tens of Thousands* of thanks.

1

SAS Can Produce High-Quality Figures

1.1 Introduction

Many people consider SAS a poor tool for producing high-quality figures. This is a misconception. However, to produce high-quality figures that meet publication requirements, it is very important that you know how to choose the right formats and fonts when using SAS/GRAPH and ODS Graphics procedures. We all have seen poor-quality figures produced using SAS in publications, including various SAS user group proceedings or even in some SAS books.

This chapter compares poor and good-quality figures produced in SAS, and discusses vector and bitmap format graphics. The chapter also discusses guidelines for electronic art requirements from some prestigious publication companies and scientific journals, including the Taylor & Francis Group's *Nature* and *Science* journals.

1.2 Two Sample Figures Produced in SAS with Different Quality

In SAS, figures can be produced in two general formats—the vector and the bitmap formats. Figures in bitmap format lose quality when they are resized, especially when the figure is not produced with enough dots per inch (DPI). and is not preferred to be used for publication purposes.

Figure 1.1 is produced using PROC GPLOT in portable network graphic (PNG) format with 75 DPI, while Figure 1.2, the same as Figure 1.1, is produced in EMF format. They might not look very different in quality on the computer screen, because the maximum screen resolution is 72 DPI, but the quality differs when printed in hard copy or enlarged on the computer screen.

Figure 1.3 shows an enlarged part of Figure 1.1. You can see that the quality is much reduced. If figures are not produced using the right formats and fonts initially in SAS, it will be difficult to *fix* the image later on. No production process can improve unclear, smudged, bit-mapped, or poorly labeled figures.

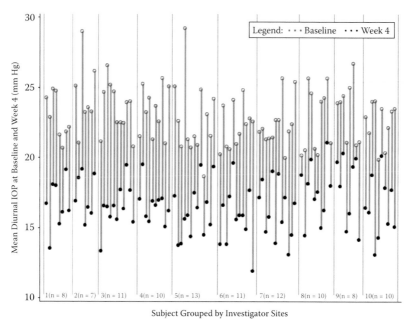

FIGURE 1.1
A sample figure in PNG format with 75 DPI.

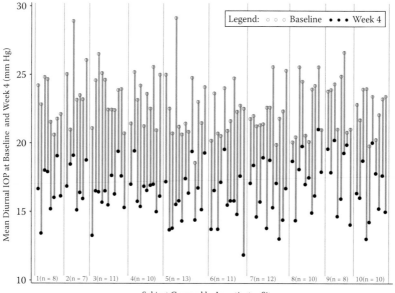

FIGURE 1.2
A sample figure in EMF format.

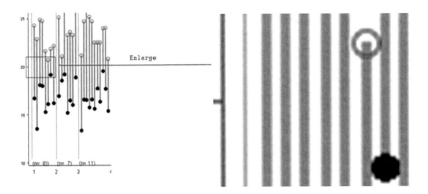

FIGURE 1.3
Enlarged bitmap format figures; quality lost when resizing.

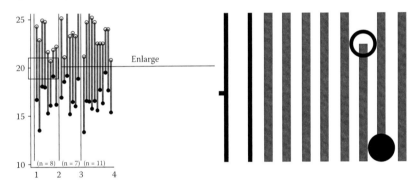

FIGURE 1.4
Vector format figures; quality preserved when resizing.

Figure 1.4 shows an enlarged part of Figure 1.2. The figure quality is preserved when resized (right side).

1.3 Bitmap and Vector Format Figures

SAS/GRAPH and ODS GRAPHICS can produce figures in two format categories—vector format and bitmap format figures (SAS Institute, 2012). The two formats differ in the way they are produced, and the quality of a vector-format figure is higher than that of a bitmap-format figure (Table 1.1).

1.3.1 Bitmap Format Graphics

A *bitmap* is a collection of bits that describe the individual pixels within an image; each pixel is a specific color, and the matrix of these pixels comprises

TABLE 1.1

Comparisons of Bitmap and Vector Format Graphics

Features	Bitmap	Vector
Plot types	JPEG, PNG, TIFF, GIF, BMP.	WMF, EMF, CGM, PS, EPS.
Production	By a collection of bits that describes the individual pixels within an image.	By direct commands with graphic descriptions, like drawing a line, filling a circle, or displaying text, etc.
Size/Resolutions	Controlled by DPI. At least 300 DPI is preferred.	Dimensionless; DPI does not apply.
Quality/Resize	Poor if DPI is low. Resizing will reduce quality.	High-quality; resizing will not reduce quality.
Applications	Better for scenes containing multiple intersecting surfaces with various shading, transparency, and lighting definitions.	Better for general publication purposes.

the image. Bitmaps can be edited only by altering individual pixels. Resizing will decrease the quality (Figure 1.3). The following are some commonly used bitmap figure formats in SAS:

JPEG: Joint Photographic Experts Group file interchange format

PNG: Portable Network Graphic

TIFF: Tagged Image File Format

GIF: Graphics Interchange Format

BMP: Windows Bitmap

1.3.2 Vector Format Graphics

Vector format figures are produced by using direct commands with graphic descriptions, like drawing a line, filling a circle, or displaying text. A vector format has no *size* or pixels, unlike the bitmap format, and is thus dimensionless. Vector format figures can be resized (shrink as small or grow as big as you want) with no change in resolution at all; the dots per inch (DPI) parameter does not apply to vector format figures. Figure 1.2, produced using the vector format, is composed of multiple individual lines and circles that are defined mathematically. When part of the figure is enlarged, the overall quality is still preserved (Figure 1.4).

The following are some commonly used vector figure formats in SAS.

WMF: Windows Metafile

EMF: Enhanced Window Metafile

CGM: Computer Graphics Metafile

PS: PostScript

EPS: Encapsulated PostScript

1.4 Why Vector Format Figures?

In general, scenes containing multiple, intersecting surfaces with various shading, transparency, and lighting definitions are displayed with greater accuracy in a bitmap format than a vector format. Vector output may differ dramatically from bitmap output, and may also differ between the vector file formats (EPS and EMF). In SAS/GRAPH, a vector format typically provides a better-quality image than a bitmap format, although with some bitmap formats you can improve the quality of the image by increasing the resolution (SAS Institute, 2012).

A bit-mapped file forms an image as a pattern of pixels (square dots) and is limited in resolution (sharpness) to the maximum resolution of the screen on which it is displayed. Bit-mapped images are inferior to vector graphics for most applications because they tend to have aliasing (also called *jaggies* and *stair stepping*), which causes a staircase distortion due to the square shapes of the pixels. Enlarging bit-mapped images accentuates the distortion and jagged edges.

A bit-mapped graphic is stored as a group of bits that represent an image to be displayed on a computer screen. The image on the screen is composed of pixels (dots), similar to the dots in a photograph in a newspaper. Each bit in an image corresponds to one pixel in the screen, so the number of pixels that composes a monitor image determines the quality of the image. Because monitor screen resolution is only 72 DPI, and the resolution needed for printing is 266 DPI, a bit-mapped image limited to 72 DPI cannot be used to produce a quality image for printing (Taylor & Francis).

1.5 Publication Requirements for Art and Figure Preparation

Although the artwork requirements of different journals and other publications, including graphs or figures, might be different, one general requirement is that the quality of figures will not be lost when the size is enlarged. Vector format figures are usually preferred to bitmapped versions because vector format figures do not have bits or dots, are dimensionless, and the quality is not lost when resizing.

The artwork guidelines from one publication company (Taylor & Francis), and two scientific journals (*Nature* and *Science*) are discussed in the following sections.

1.5.1 Taylor & Francis Group: Electronic Art Guidelines

Taylor & Francis strongly prefers vector format graphics and discourages the submission of bitmap format graphics. Below are the guidelines on both vector and bitmap format graphics (Taylor & Francis Group).

1.5.1.1 Vector Graphics Formats

A vector file creates an image as a collection of lines rather than as a pattern of individual pixels (bitmapped graphics). Vector files are much easier to edit than bitmapped graphics (objects can be individually selected, sized, moved, and otherwise manipulated) and are preferred for professional illustration purposes. Because they are scale- and resolution-independent, vector images can be enlarged without loss of sharpness. The following are acceptable vector file formats in order of preference:

- *Adobe Illustrator (AI):* This is the vector graphics program that is best suited for creating high-quality professional graphics.
- *PDF:* This file format allows a document to be transferred to another type of computer system without losing the original formatting.
- *EPS:* EPS is a high-resolution graphic image stored in the PostScript language. The EPS format allows users to transfer high-resolution graphics images between applications. The images can also be sized without sacrificing quality.

1.5.1.2 Bitmapped Graphics Formats

Although their use is discouraged, the following bitmapped graphics formats are listed in order of preference:

- *GIF:* GIF was developed to exchange graphics files over the Internet. Although GIF files are widely used, the JPEG format reduces graphics files to about one-third the size of a GIF file, leading to faster Internet transmission. GIF files are more efficient than JPEG files if an image contains many solid areas.
- *JPEG:* JPEG is a graphics format that is specifically designed for photographic images and other complex pictures such as realistic artwork. It is not well suited to line drawings, text, or simple cartoon illustrations.
- *TIFF:* TIFF is a bitmapped graphics format commonly used for the scanning, storage, and interchange of grayscale graphic images.

1.5.2 *Nature*: Final Artwork

When a manuscript is accepted for publication, the author is required to submit high-resolution figures. Some of the acceptable formats include: AI, EPS, PS, and PDF (*Nature Guide to Authors*).

1.5.3 *Science*: Preparing Your Art and Figures

The *Science* journal has very strict requirements for figure resolution and format at the manuscript revision stage (*Science*, "Preparing Your Art and Figures").

Resolution. For manuscripts in the revision stage, adequate figure resolution is *essential* to a high-quality print and online rendering of your paper. Raster line art should carry an absolute minimum resolution of 600 DPI and, preferably, should have a resolution of 1200 DPI. Grayscale and color artwork should have a minimum resolution of 400 DPI, and a higher resolution if possible.

Please note that these resolutions apply to figures sized at dimensions comparable to those of figures in the print journal. Reducing or enlarging the dimensions of a digital raster image will also change its resolution. For example, reducing the dimensions of an image by 50%, with no change in file size, will double its DPI resolution; doubling the dimensions of the image will cut resolution by 50%. Authors are encouraged to review past issues to gauge the approximate size their figures will take in the print publication, and set the resolution of their figures accordingly.

Format. Electronic figure files at the revision stage must be in one of the following formats: PDF, PS, or EPS for illustrations or diagrams; TIFF, EPS, PS, or PDF for photography or microscopy. Authors who have created their files using Adobe Illustrator or Adobe Photoshop should provide their files in these native file formats.

1.6 SAS Can Produce High-Quality Figures to Meet Publication Requirements

The vector and high-resolution bitmap format figures required for publication purposes by publication companies and scientific journals can be produced in SAS. The following chapters of the book discuss the principles and include practical examples and SAS programs to illustrate how to produce different types of high-quality figures using procedures in both SAS/GRAPH and ODS Graphics systems. A reader can easily learn and modify the SAS programs that are included in this book to produce their own custom-made figures.

1.7 References

Nature Publishing Group. "Guide to Preparing Final Print-Only Artwork." In *Nature Guide to Authors: Final Artwork*, Nature Publishing Group, n.d., http://www.nature.com/nature/authors/gta/3c_Final_artwork.pdf.

SAS Institute Inc. 2012. *SAS/GRAPH® 9.3: Reference,* 3rd ed. Cary, NC: SAS Institute Inc.

SAS Institute Inc. 2012. *SAS® 9.3 ODS Graphics: Procedures Guide,* 3rd ed. Cary, NC: SAS Institute Inc.

Science, "Preparing Your Art and Figures," *Science* magazine, n.d., http://www.sciencemag.org/site/feature/contribinfo/prep/prep_revfigs.xhtml.

Taylor & Francis Group. "Author's Guide to Publishing: Disk Manuscripts." http://www.eng.auburn.edu/~wilambm/monographs/inf/TF%20guidelines_disk%20book_May2008.pdf

2

Principles of Producing High-Quality Figures in SAS

2.1 Introduction

In Chapter 1, we discussed the vector and bitmap format figures and why vector format graphics are preferred to bitmap format for publication purposes. Vector format graphics are preferred because they are of high quality and can be enlarged without loss of sharpness. In this chapter, we will discuss some major principles to produce high-quality figures in vector or high-resolution bitmap formats. In statistical analysis software (SAS), figures can be produced in listing outputs (PS, EPS, EMF, PNG, etc.) and ODS outputs (PDF, RTF, etc.) using both SAS/GRAPH and ODS Graphics procedures (SAS Institute Inc., 2012).

2.2 SAS/GRAPH and ODS Graphics Procedures

There are two very distinct systems to produce graphics in SAS—SAS/GRAPH and ODS Graphics. SAS/GRAPH produces graphics using a device-based system, while ODS Graphics produces graphics using a template-based system through the SAS Output Delivery System (ODS) (SAS Institute Inc., 2012).

2.2.1 SAS/GRAPH Procedures

SAS/GRAPH procedures that produce device-based graphics include GCHART, GPLOT, GMAP, GBARLINE, GCONTOUR, and G3D procedures. For device-based graphics, the GOPTIONS statement is used to control the graphical environment. For example, the DEVICE = option in the GOPTIONS statement is used to generate SAS/GRAPH output in specified formats. Section 2.5 describes, in detail, the GOPTIONS used in this book.

2.2.2 SAS ODS Graphics Procedures

The ODS Graphics procedures that produce template-based graphics include SGPLOT, SGPANEL, SGSCATTER, SGDESIGN, and SGRENDER. Device drivers and most SAS/GRAPH global statements (AXIS, LEGEND, PATTERN, and SYMBOL, etc.) have no effect on template-based graphics. You must use the ODS GRAPHICS statement to control the graphical environment. For example, you can select the type of image file (PS, EMF, GIF, PNG, JPEG, etc.) by specifying the OUTPUTFMT = option in the ODS GRAPHICS statement.

The ODS Graphics procedures were previously part of SAS/GRAPH and were called SAS/GRAPH Statistical Graphics (SG) procedures in SAS 9.2. However, they have been moved to Base SAS in SAS 9.3. SAS/GRAPH software is no longer required in order to use the ODS Graphics procedures. The ODS Graphics Designer, ODS Graphics Editor, and Graph Template Language have also been moved to Base SAS.

2.2.3 SAS/GRAPH versus ODS Graphics Procedures

Although the SAS Institute has shifted its resources to developing and updating the ODS Graphics procedures, it will always keep SAS/GRAPH in the software and make it available. While the ODS Graphics procedures provide a more convenient and efficient way to produce some types of figures, especially for multicell or multipanel figures, the traditional procedures of SAS/GRAPH are still very versatile and might be preferred to produce some custom, single-cell figures because they provide very high flexibility to control the appearance of the output. Table 2.1 summarizes the main differences between the SAS/GRAPH and the ODS Graphics procedures (SAS Institute Inc., 2012; Zender and Kalt, 2012). More detailed discussion of the differences between the two systems in producing different types of figures is discussed in the corresponding chapters in this book.

Because there are two quite distinctive systems to produce SAS graphics, you might wonder which one you should use. Experienced SAS programmers and statisticians who have accumulated many SAS/GRAPH programs to generate various types of figures might stay with SAS/GRAPH. New or novice programmers and statisticians might choose to go with the relatively easier ODS Graphics. Zender and Kalt at the SAS Institute have a very good suggestion: "You might stay on one graph road for your whole career as a SAS user. But you might, at some point, need to switch over to the other road, or better yet, spend some time on both roads" (Zender and Kalt, 2012).

Almost all sample figures (except a few) in the book are produced using both the SAS/GRAPH and ODS Graphics procedures and the pros and cons in using both procedures are compared and discussed in the corresponding chapters.

TABLE 2.1

Main Differences between SAS/GRAPH and ODS Graphics Procedures

Features	SAS/GRAPH	ODS Graphics
Graph size, output formats, and name	Controlled by GOPTIONS statement with options such as HSIZE =, VSIZE =, GSFNAME =, and DEVICE = options	Controlled by the ODS GRAPHICS statement with options such as HEIGHT =, WIDTH =, OUTPUTFMT =, and IMAGENAME = options
Properties for text, markers, and lines, etc.	Global statements such as AXIS, LEGEND, and SYMBOL are used to set the properties for text, markers, and lines.	Statements or options within the procedure, such as LINEATTRS and MARKERATTRS, are used to control visual properties.
Plot types	Determined by global options. The INTERPOL = option in the SYMBOL statement determines whether a graph is a scatter plot or a box plot in some graphs.	Determined by the plot statement only, such as SERIES, SCATTER statements in SGPLOT
Fonts (software, hardware, and system)	Supports software (such as SWISSB) and hardware fonts, including the device-resident (such as PostScript) and system fonts (such as Arial)	Only supports system fonts (such as Arial)
RUN-group	Some procedures, such as GPLOT, support RUN-group processing to produce multiple figures within the same procedure.	RUN-group processing is not supported.
Annotation	The SAS/GRAPH Annotate facility is supported.	SG Annotation facility is supported.
Multiple graphs on a page	Using PROC GREPLAY	Using SGPANEL or SGSCATTER procedures, or GTL
Flexibility in controlling the figure appearance	Very high degree	Moderately high degree.

2.3 Listing Output to Save Figures from One Procedure (PS, EPS, EMF, CGM, and so on)

The commonly used listing output formats include PS, EPS, WMF, EMF, CGM, PNG, JPEG, and TIFF, and can be produced with both SAS/GRAPH and ODS Graphics procedures.

If SAS/GRAPH procedures are adopted to produce the listing outputs, it is important that the right device drivers are chosen by specifying the DEVICE = option in the GOPTIONS statement. Table 2.2 lists the suggested device drivers for some commonly used listing output formats (SAS Institute Inc., 2005).

TABLE 2.2

Graphic Output Formats and Device Drivers

Output Format	Suggested Device Driver
EPS, PS	PSLEPSFC or PSCOLOR (color)
	PSLEPSF (monochrome)
EMF	SASEMF
WMF	SASWMF
CGM	CGMOFML (landscape)
	CGMOFMP (portrait)
PNG	PNG
JPEG	JPEG
TIFF	TIFFP (color)
	TIFFB (monochrome)

When using the FILENAME statement together with the GSFNAME = option in the GOPTIONS statement to save listing format files, it is important that you turn on the ODS Listing, otherwise the listing file might not be saved.

```
** Produce listing outputs (PS, EMF);
ods listing;
```

If ODS Graphics procedures are used to produce these listing format output files, the format can be specified in the *OUTPUTFMT = Option* in ODS Graphics statement.

```
ods graphics/reset = all width = 8in height = 6in OUTPUTFMT = EMF;
```

2.4 ODS Output to Save Figures from Several Procedures into a Document (PDF, RTF, and so on)

The listing output format files in Section 2.3 can only save the figures produced by the same procedure. To save multiple figures produced by multiple SAS/GRAPH and ODS Graphics procedures in one document file, you can use ODS output statements, such as ODS PDF or ODS RTF, to save in PDF or RTF format document files. The RTF format file is designed for sharing documents among Microsoft applications, including Word and PowerPoint. Figures in the RTF file can easily be copied and resized without quality loss. A PDF file is usually preferred for the purpose of publication. The RTF format figures are referred to as *in-text figures* while the PDF format figures are referred to as *post-text figures*.

The ODS RTF format file can store EMF, JPEG, PNG, or ACTIVEX format figures, and the SASEMF driver, which produces EMF format figures, is the default device driver. SASEMF, PNG, JPEG, and ACTIVEX are the recommended device drivers for ODS RTF (SAS Institute Inc., 2005). A relatively new device, PNG300, can be used to save high-resolution PNG-format figures with 300 dots per inch (DPI) in an RTF file.

PDF format can store multiple page postscript format figures produced from several procedures, and SASPRTC is the default device driver with SASPRTC (color), SASPRTG (gray scale), and SASPRTM (monochrome) as the recommended device drivers (SAS Institute Inc., 2004). If the figures are produced in one procedure, including multiple figures by using a BY statement or RUN-group process in SAS/GRAPH, the native PDF device drivers PDF and PDF C can be used.

When producing ODS output files, it is suggested that ODS LISTING be turned off.

```
** Produce ODS outputs (PDF and RTF);
ods listing close;
ods pdf file = "&outloc./Figures in PDF File.pdf" nogtitle
nogfootnote;
** SAS/GRAPH or ODS/Graphics Procedures producing figures;
ods pdf close;

ods rtf file = "&outloc./Figures in RTF File.rtf" nogtitle
nogfootnote;
** SAS/GRAPH or ODS/Graphics Procedures producing figures;
ods rtf close;
```

PDF files can also be created indirectly by combining the postscript listing files (PS and EPS) into one PDF file manually or using various utilities.

Besides PDF and RTF files, you can also produce HTML (HyperText Markup Language) documents with GIF, JPEG, PNG, or ACTIVEX graphs.

2.5 GOPTIONS in SAS/GRAPH

A GOPTIONS statement is required in SAS/GRAPH procedures to select the image format, layout, and visualization. Like SAS system OPTIONs, the scope is global. GOPTIONS only applies to SAS/GRAPH and does not have any influence on SAS Graphic procedures, like SGPLOT and SGPANEL.

The following is the GOPTIONS statement that is used in the SAS/GRAPH procedures (GPLOT, GCHART, etc.) that are included in this book.

```
goptions
   reset    = all
   GUNIT    = PCT
   rotate   = landscape
   gsfmode  = replace
   gsfname  = GSASFILE
   device   = &DRIVER
   lfactor  = 1
   hsize    = 8 in
   horigin  = 0 in
   vsize    = 6.5 in
   vorigin  = 0 in
   ftext    = "&FONTNAME"
   htext    = 10pt
   ftitle   = "&FONTNAME"
   htitle   = 10pt
;
```

The following are descriptions of each option included, and is based on "Graphics Options and Device Parameters Dictionary" in the SAS/GRAPH 9.3 Reference (SAS Institute Inc., 2012).

RESET = *ALL* | *GLOBAL* | statement-name(s): Option ALL sets all graphics options to defaults and cancels all global statements, GLOBAL cancels all global statements (AXIS, FOOTNOTE, LEGEND, PATTERN, SYMBOL, and TITLE), and "statement-name(s)" resets or cancels only the specified global statements. To cancel several statements at one time, enclose the statement names in parentheses. For example, RESET = (TITLE FOOTNOTE).

GUNIT = units: This option specifies the default unit of measurement used with height specifications. Choices include CELLS (character cells), CM (centimeters), IN (inches), PCT (percentage of graphics output area), and PT (points; there are about 72 points in an inch).

ROTATE = : This option can be landscape or portrait. Landscape is usually the preferred choice.

GSFMODE = *APPEND* | *PORT* | *REPLACE*: This option specifies the disposition of records that are written to a graphics stream file (GSF) or to a device or communications port by the device driver. REPLACE replaces the existing contents of a GSF that is designated by the GACCESS = or GSFNAME = graphics option or device parameter. If the file does not exist, it is created. REPLACE is always the default, regardless of the destination of the GSF. APPEND adds the records to the end of a GSF that is designated by the GACCESS = or GSFNAME = graphics option or device parameter. If the file does not already exist, it is created. PORT sends the records to a device or communications port. The GACCESS = graphics option or device parameter should point to the desired port or device.

DEVICE = device-entry: Specifies the device driver to which SAS/GRAPH sends the procedure output. The device driver controls the format of graphics output. Different device drivers are used to produce figures in different formats. In this book, "PSCOLOR" is used to produce PS or EPS format figure and SASEMF is used to produce figures in an EMF format. SASEMF is also the default device for the ODS RTF destination. For ODS PDF output, the device driver is SASPRTC.

GSFNAME = fileref: This option specifies a fileref that points to the destination for the GSF output. Fileref must be a valid SAS fileref up to eight characters long and must be assigned with a FILENAME statement before running a SAS/GRAPH procedure that uses that fileref.

LFACTOR = line-thickness-factor. This option can range from 0 through 9999. A value of 0 for LFACTOR is the same as a factor of 1. A *line-thickness-factor* value of 2, for example, causes the line to be two times as thick as normal. When using GPLOT to produce line plots in PS format or in PDF files, LFACTOR = 2 is better than 1, when producing EMF format or RTF files, LFACTOR = 1 is acceptable.

FTEXT = "&FONTNAME": This option sets the font for all text in the graph image. A macro variable *&FONTNAME* is used in this book to produce figures with the selected fonts. Depending on the device driver (PSCOLOR, SASEMF etc.), the hardware fonts, like Times or Courier New, can be used.

VSIZE = 8 in and HSIZE = 6 in: This option sets the default size in inches for the vertical and the horizontal axes. For landscape layout figures, a ratio of about 4:3 for the vertical versus the horizontal size is suggested. Because the HSIZE = and VSIZE = options are specified, GUNIT is ignored and inches will be used for the vertical and horizontal axes.

HTEXT = 10pt and HTITLE = 10pt: This option sets the default text and TITLE1 heights.

Other SAS/GRAPH graphics options might not be as commonly used and are not discussed here. A complete list of graphics options can be found in SAS/GRAPH 9.3: Reference (SAS Institute Inc., 2012).

If we want to produce bitmap format figures with high resolution, we can do that by setting the XMAX, YMAX, XPIXELS, and YPIXELS options on the GOPTIONS statement. The following example will produce a PNG format figure with a resolution of 300 DPI. The number in XPIXELS divided by the number in XMAX (XPIXELS/XMAX) is the DPI number and should be the same as YPIXELS/YMAX.

```
goptions dev = png xpixels = 1200 ypixels = 900 xmax = 4 in
ymax = 3 in;
```

In ODS Graphics procedures, like SGPLOT, the high-quality bitmap format figures can be produced by specifying the DPI using the statement *ods listing image_dpi = 300.*

2.6 Software vs. Hardware Fonts

2.6.1 Software Fonts

Software fonts are those created by SAS Institute, and are saved in SASHELP. FONTS. By default, SAS/GRAPH searches for the fonts in SASHELP.FONTS category. SAS software fonts, such as SWISS, ZAPF, and CENTX, can be used to create texts in all the graphics formats in SAS/GRAPH. SAS software fonts can be used in codes and output when exporting graphics on different operating systems and with different device drivers (SAS Institute Inc., 2005).

2.6.2 Hardware Fonts

In SAS/GRAPH, hardware fonts are native to the graphics format, and the availability depends on the device driver and operating systems. Some commonly used hardware font sets include system fonts and a standard subset of PostScript fonts. Hardware fonts are independent of the software/printer used and can be used to produce figures of high quality. Hardware fonts often produce clearer results, with smaller output files that can be produced faster.

The hardware fonts that are available to a particular device driver are named in the Chartype list of that driver. To display this list, submit the following code:

```
proc gdevice c = sashelp.devices nofs;
   list DeviceDriverName;
run;quit;
```

Replace *DeviceDriverName* with the name of the device driver you want listed. For example, if SASEMF is selected and run, the following hardware fonts will be displayed in the Chartype list.

```
Chartype    Font Name
   0        <MTmonospace>
   1        <MTserif>
   2        <MTsans-serif>
```

If PSCOLOR is selected, only the DMS font is listed in the Chartype list. For device drivers that explicitly list fonts by name, the exact names should be used to reference the fonts. For device drivers that only list "DMS Font," the system fonts can be used.

Window System Fonts: When SAS is run on a Windows system, those device drivers that list the "DMS Font" will recognize the Windows system fonts, which are called True Type fonts. These are the same fonts that are used in Microsoft Office. To see a list of available True Type fonts, do the following:

- Use the font browser in the Microsoft Office application.
- In SAS, select the **File** menu, choose **Print Setup**, and then click the **Fonts** button. Any font with the Text icon (two overlapping capital T's) next to its name is a True Type font.

True Type font names are case-sensitive, so they must be referenced exactly as shown in this list.

2.6.3 Available Hardware Fonts for Listing Outputs

The following are the hardware fonts that are associated with the PSLEPSF and PSLEPSFC (or PSCOLOR) and SASEMF device drivers, which can be used in SAS/GRAPH to produce high-quality listing output files (PS, EPS, EMF, etc.). In ODS Graphics, only system fonts are supported.

PSLEPSF and PSLEPSFC (or PSCOLOR): Commonly used fonts are Courier, Courier-Oblique, Courier-Bold, Courier-BoldOblique, Times-Roman, Times-Italic, Times-Bold, Times-BoldItalic, Helvetica, Helvetica-Oblique, Helvetica-Bold, Helvetica-BoldOblique, and Symbol, etc. The complete list can be found in SAS TS-674 (SAS Institute Inc., 2005).

It is recommended that a hardware font, one native to the PDF format, be used to generate the output. Text that is created in a font that is also available to the program that is viewing the file will be rendered in the correct font. Note that most publishers that accept postscript files will also require that you use one of the standard set of Adobe PostScript fonts in your figure. The required fonts are typically either Helvetica or Times, or Symbol for Greek fonts.

SASEMF: The SASEMF device driver uses this subset of fonts on all systems: Courier New, Courier, and Letter Gothic.

With EMF, WMF, and CGM graphics, text created in a hardware font will be editable after the graph is imported into Microsoft Office. With the other formats, none of the text will be editable, but text in a hardware font may appear crisper in the output.

2.6.4 Available Hardware Fonts for ODS Outputs (RTF, PDF)

The document formats RTF and HTML generally support system fonts for text outside the graphs, such as tables, titles, or footnotes. Texts within the graphs that are stored in or with these documents are limited to the fonts that are supported by the device driver that is used to create the graph.

RTF File (with SASEMF Device): Text or tabular output written to an RTF file can use True Type or PostScript fonts, depending on the operating system. For EMF graphics, the SASEMF device driver is used by default. Text stored in this graphics format is restricted to the Courier, Courier New, and Letter Gothic fonts. Specifying the PNG or JPEG device drivers will make system fonts available for the graphs. To produce text in a particular font, it may be necessary to set both the DEVICE and TARGET to the same device driver (SAS Institute Inc., 2005).

PDF File (with Universal Printer Device SASPRTC): For the PDF universal printer, it is recommended that the Base 14 fonts that are installed by default with the Adobe Acrobat Reader be used (SAS Institute Inc., 2004). These fonts are referenced in Table 2.3.

2.7 Controlling Titles and Footnotes in RTF Format Figure Files

When using ODS RTF together with graph procedures in SAS/GRAPH and ODS Graphics, ODS RTF places the titles within the boundary of the figure image by default. You can choose the placement of the titles and footnotes by specifying the options in ODS RTF.

TABLE 2.3

Names for the Base 14 Fonts with the PDF Universal Printer Using SASPRTC Device

Courier	Helvetica	Times	Symbol
Courier/Oblique	Helvetica/Oblique	Times/Italic	ITC Zapf Dingbats
Courier/Bold	Helvetica/Bold	Times/Bold	
Courier/Bold/Oblique	Helvetica/Bold/Oblique	Times/Bold/Italic	

Note: Font names are not case-sensitive, but they must be referenced in quotes.

- Inside the image (default): GTITLE, GFOOTNOTE
- Outside the image in headers/footers: NOGTITLE, NOGFOOTNOTE
- Outside the image but not in headers/footers: NOGTITLE, NOGFOOTNOTE, and BODYTITLE or BODYTITLE_AUX

Most of the time, you may only need to copy and paste the graphics image itself into Word applications and write the titles and footnotes directly in the document. The NOGTITLE and NOGFOOTNOTE options place the titles and footnotes in the header and footer sections of the RTF document, and are used in this book to produce RTF format figure files.

2.8 Adding Bookmarks in PDF Format Figure Files.

In SAS/GRAPH, the ODS PROCLABEL = statement can be used to add bookmarks in graphs in the ODS PDF destination. The PROCLABEL = option specifies the name of the top-level bookmark. The description for each procedure that is run after the ODS PROCLABEL = statement is added as a subtopic under the top-level bookmark that the PROCLABEL = option defines. The DESCRIPTION = option is used to set the text of the subtopic bookmark for each graph procedure. Figure 2.1 displays an example of a PDF file with bookmarks produced in SAS.

In ODS Graphics, however, the DESCRIPTION = option does not work and you can only specify one level of bookmarking using PROCLABEL = option.

2.9 Setting Up Global Options and Macro Variables

The SAS options and macro variables *pgmname, pgmloc, pgmpth,* and *outloc* are defined at the beginning of every SAS program in this book to set up

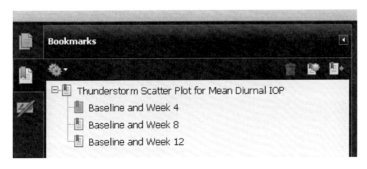

FIGURE 2.1
Bookmarks in a PDF file produced in GPLOT.

the global options, the program name, the location, the path, and the graph output location.

```
options mprint symbolgen nodate nonumber validvarname = v7
  orientation = landscape;
%let pgmname = Line Plots.sas;
%let pgmloc = C:\SASBook\SAS Programs\Chapter 3;
%let pgmpth = &pgmloc.\&pgmname. &sysdate9. &systime. SAS
V&sysver.;
%let outloc = C:\SASBook\Sample Figures\Chapter 3;
```

The options *mprint, symbolgen* are used for debugging purposes. These options put the macro information and the macro symbol–generated info in the SAS LOG window. The options *nodate, nonumber* are used because date and page number are normally not needed in figures. The date and time running the SAS programs are included in the *pgmpth* macro variable to be placed in the footnotes by using the system macros *&sysdate.* and *&systime.* Landscape is the preferred page layout for figures.

The option *validvarname = v7* is selected to allow only valid variable names based on SAS version 7. This option is useful when running the programs in the SAS Enterprise Guide (EG) environment, because we do not want to allow empty space in variable names (by default, SAS EG would allow empty space between characters in a variable name). If the option is not specified, some programs that work well in PC SAS might not be run successfully in SAS EG and might end up with ERROR messages in LOG, especially if PROC TRANSPOSE is used.

2.10 References

SAS Institute Inc. 2004. "Exporting SAS/GRAPH Output to PDF Files from Release 8.2 and Higher." SAS Knowledge Base paper TS-659, http://support.sas.com/techsup/technote/ts659/ts659.html#B3b.

SAS Institute Inc. 2005. "An Introduction to Exporting SAS/Graph Output to Microsoft Office SAS Release 8.2 and Higher." SAS Knowledge Base paper TS-674, http://support.sas.com/techsup/technote/ts674/ts674.html#IIC1.

SAS Institute Inc. 2012. "Differences between the ODS Graphics Procedures and SAS/GRAPH Procedures." In *SAS® 9.3 ODS Graphics: Procedures Guide*, 3rd ed. Cary, NC: SAS Institute Inc.

SAS Institute Inc. 2012. *SAS/GRAPH® 9.3: Reference*, 3rd ed. Cary, NC: SAS Institute Inc.

Zender, C., and Kalt, M. 2012. "At the Crossroads: How to Decide on Your Graphics Path." In *SAS Global Forum 2010 Proceedings*. Cary, NC: SAS Institute Inc.

3

Line Plots

3.1 Introduction

Line plots, often used to display the overall mean response pattern of subjects in a sample by time, is one of the most commonly used plots in scientific research and publications. This chapter illustrates how to produce high-quality line plots, including simple line plots with only mean values by time, and more complicated line plots with standard deviation (SD) or maximum and minimum value bars displayed together with the mean values.

The same line plots are produced using both PROC GLOT in SAS/GRAPH and PROC SGPLOT in ODS Graphics. The features, pros, and cons associated with the two procedures in producing line plots are discussed. The SAS programs producing the sample figures are discussed and included in the Appendix (Section 3.6).

3.2 Application Examples

To illustrate the application and production of line plots, four sample figures (Figures 3.1 to 3.4) are presented in the chapter. The basic concepts and techniques in producing the line plots described in the chapter can easily be applied to other clinical and nonclinical research areas.

Sample Figures 3.1 and 3.2 are based on a virtual clinical trial to compare the effects of a New Drug with that of an Active Control on intraocular pressure (IOP) reduction in patients with glaucoma or ocular hypertension (OHT). The IOP values are measured three times within a day at 8 a.m., 10 a.m., and 4 p.m. of the baseline and weeks 4, 8, and 12 visits. There are 500 patients enrolled at 10 investigator sites. The subjects are randomized to receive either the New Drug or the Active Control at the ratio of 1:1. Figure 3.1 is a simple line plot displaying the mean IOP values for the two treatment groups over time. Figure 3.2 is a more complicated line plot than Figure 3.1 with SD bars displayed together with the mean IOP values at each hour overtime.

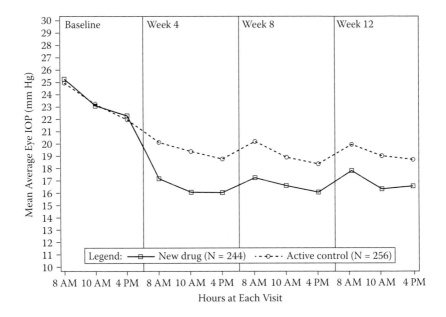

FIGURE 3.1
Mean IOP by treatment at each visit/hour; a simple line plot.

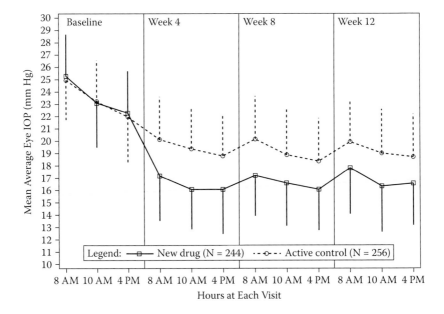

FIGURE 3.2
Mean IOP by treatment at each visit/hour; a line plot with SD bar displayed.

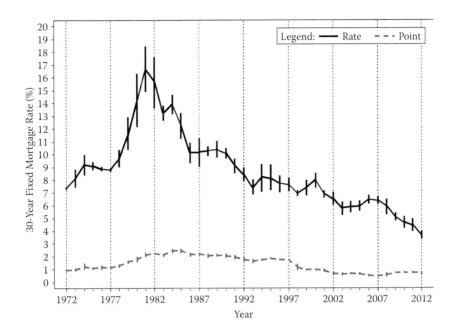

FIGURE 3.3
Maximum, average, and minimum rates and points for a 30-year fixed-rate mortgage; 1972 to 2002 with rates and points displayed in the same y-axis.

Sample Figures 3.3 and 3.4 are based on historical mortgage data from Freddie Mac (Freddie Mac, 2013). Figure 3.3 is a line plot with the maximum, average, and minimum rates and points for 30-year fixed-rate mortgages from 1972 to 2012. Figure 3.4 is the same as Figure 3.3 except that the mortgage rates and points are positioned in two separate y-axes.

3.2.1 A Simple Line Plot with Mean IOP Values by Time

A line plot displaying the mean IOP value at each hour of each visit is a good data visualization tool to examine the effects of the two drugs on the mean IOP reduction by time. Figure 3.1 demonstrates that the subjects in the two groups have almost identical mean IOP values at the 3 hours of the baseline visit. After treatment, subjects in the New Drug treatment group have consistently lower mean IOP values than those in the Active Control at all post-baseline visits and hours, indicating a better IOP reduction effect for New Drug.

3.2.2 A Line Plot with SD Bar Displayed Together with the Mean IOP Values

Besides the mean values, you may also want to examine the data variation by the SD values at each visit and hour. Figure 3.2 displays the SD bar at each

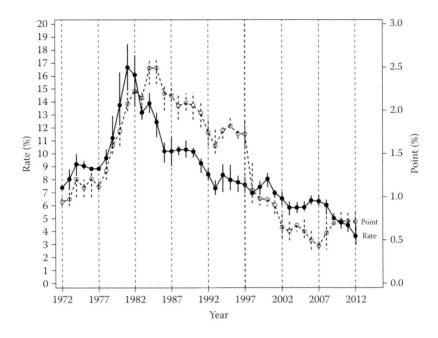

FIGURE 3.4

Maximum, average, and minimum rates and points for a 30-year fixed-rate mortgage; 1972 to 2002 with rates and points displayed in the different y-axes.

hour together with the mean values, and it seems that the data variations between the two treatment groups are quite consistent. Please note that the SD bar at the same hour for the two treatment groups are positioned in different directions based on the mean values. If the mean value is higher, it is displayed in the upper position, otherwise in the lower position.

3.2.3 Line Plots with Maximum, Average, and Minimum Rates and Points for a 30-year Fixed-Rate Mortgage

Freddie Mac keeps the historical primary mortgage market survey (PMMS) data online. The 30-year and 15-year fixed-rate, and the 5-year and 1-year adjustable-rate PMMS data are available for free download online (Freddie Mac, 2013). Figures 3.3 and 3.4 display the average together with the maximum and the minimum rates and points of the 30-year fixed-rate mortgage from 1972 to 2012. Figure 3.3 displays both the rates and points in the same y-axis. Since points are much smaller than the rates and might not be distinguished well among years when put in the same axis as the rates, Figure 3.4 is produced to position the rates and points in two different y-axes. From the line plots, we can see that the highest rate in a 30-year mortgage occurred in 1981 with the average rate greater than 16% and the maximum rate close to 19%. The highest mortgage points occurred in the years of 1984 and 1985 (close to 2.5%), not in the same year when the rates are the highest (1981).

Readers can easily modify the SAS programs included in the chapter to produce similar line plots for other mortgage products in PMMS.

3.3 Producing the Sample Figures

3.3.1 Data Structure and SAS Annotated Dataset

Table 3.1 displays the data structure for one subject's IOP data at each visit and hour. There are 500 subjects with a total of 6,000 records in the simulated dataset IOP.

In PROC GPLOT, the SAS/GRAPH annotate facility is used to place the SD bars in either the upper or the lower direction from the mean values at each visit/hour. One subject's record in the annotated dataset is shown in Table 3.2. At "time 1" (8 a.m., Baseline) SAS at first "moves" the point to the (1, 24.62) coordinate, and then "draws" a line to the (1, 21.20) coordinate in red color. The line type and size are controlled by the value of variables "line" and "size." For a more detailed introduction to the SAS annotate facility, please see "Part 5, The Annotate Facility" in the *SAS/GRAPH 9.3 Reference* (SAS Institute Inc., 2012).

3.3.2 Notes to SAS Programs

The two SAS programs used to produce the four sample line plots using both the GPLOT and the SGPLOT procedures are provided in the Appendix (Section 3.6).

TABLE 3.1

Part of the Simulated IOP Data (One Subject's Data)

subjid	site	trtnum	visit	hour	iop
1001	1	Active Control	Baseline	Hour 0	24.5
1001	1	Active Control	Baseline	Hour 2	25.3
1001	1	Active Control	Baseline	Hour 8	20.4
1001	1	Active Control	Week 4	Hour 0	15.3
1001	1	Active Control	Week 4	Hour 2	22.9
1001	1	Active Control	Week 4	Hour 8	17.3
1001	1	Active Control	Week 8	Hour 0	13.7
1001	1	Active Control	Week 8	Hour 2	20.6
1001	1	Active Control	Week 8	Hour 8	17.1
1001	1	Active Control	Week 12	Hour 0	23.5
1001	1	Active Control	Week 12	Hour 2	17.2
1001	1	Active Control	Week 12	Hour 8	16.7

TABLE 3.2

Part of the Annotated Dataset Used in GPLOT

time	trtnum	FUNCTION	COLOR	XSYS	YSYS	HSYS	SIZE	X	Y	mn	sd	LINE
1	New Drug	MOVE	red	2	2	4	2	1	24.62	24.62	3.42	
1	New Drug	DRAW	red	2	2	4	2	1	21.20	24.62	3.42	1
1	Active Control	MOVE	black	2	2	4	2	1	25.06	25.06	3.51	
1	Active Control	DRAW	black	2	2	4	2	1	28.57	25.06	3.51	2

3.3.2.1 Main Sections and Features of the First Program

3.3.2.1.1 Dataset Simulation

- Clinical trial IOP data is simulated to include 500 subjects in 10 sites. Subjects are randomly assigned to each treatment group at the ratio of 1 to 1 by using the function *"ranuni(0)"*. If the random number generated by *"ranuni(0)"* is < 0.5, then the subject is assigned to New Drug; otherwise, the subject is assigned to the Active Control group.
- Subjects' IOP values are assigned based on a normal distribution with the preset mean and SD. Subjects' IOP for the two treatment groups are the same at baseline, but different at post-baseline visits assuming the New Drug has better effects in reducing the IOP values than the Active Control. The IOP values at the 3 hours of a day are also different to mimic the natural diurnal pattern of the IOP values.

3.3.2.1.2 Data Analyses and Manipulation

- The mean IOP values at each visit and hour of each treatment group is calculated using Proc Means with the results saved into a dataset *"iop_stat"*.
- The visit and hour is assigned to a variable *"time"* with values from 1 to 12.
- The dataset with IOP mean value for the 2 treatment groups is sorted by the *"time"* variable first, then by the *"mn"* (mean value) of the two groups. This sorting makes the treatment group with the higher mean value of the same hour appear first, which is important when positioning the SD bar in the upper or lower position from the mean.

3.3.2.1.3 Annotate Dataset for Drawing the SD Bar in GPLOT

The SAS annotate macros %*ANNOMAC*,%*DCLANNO*, and %*LINE* are used to create annotate datasets to draw the SD bars at each hour. The functions and use of the three annotate macros are described in the following bullets (SAS Institute Inc., 2012).

- The %*LINE* macro draws a line between two sets of coordinates. It has the syntax of %*LINE (x1, y1, x2, y2, color, line, size)*. You must run the %*ANNOMAC* and %*DCLANNO* macros before using the %*LINE* annotate macro.
- The %*DCLANNO* annotate macro automatically sets the correct length and data type for all annotate variables except the TEXT variable. It has the syntax of %*DCLANNO* without any macro arguments.
- The %*ANNOMAC* macro must be run before any other annotate macros are used in an SAS session. After submitting the macro %*ANNOMAC*, you can see the following messages in the LOG window.

```
*** ANNOTATE macros are now available ***
   For further information on ANNOTATE macros, enter,
      %HELPANO(macroname), (for specific macros)
      %HELPANO(ALL), (for information on all macros)
   or%HELPANO (for a list of macro names)
```

- The *"Mod(_N_, 2)"* function is used to decide the SD bar direction. If the result from the function is 1 (indicating that _N_ is an odd number), the mean value of the treatment is smaller than the other treatment group, the SD bar is positioned in the lower direction; otherwise, the bar is put in the upper direction. Please note that "_N_" is a counter that SAS uses to track the iterations through the implicit loop of the DATA step.

3.3.2.1.4 The RUN-Group Feature in GPLOT

The GPLOT procedure in SAS/GRAPH supports the RUN-group feature, and multiple figures can be produced within one GPLOT procedure using the same dataset.

- A RUN statement is required to produce each figure after the PLOT statement.
- A FILENAME statement is used with each PLOT statement to save each figure.
- A "DES =" option after the plot statement is used to describe the plot. This is a good feature to use to produce figures in PDF documents for bookmarks.

```
proc gplot data = MN_IOP;
   FILENAME GSASFILE "&OUTLOC./Figure 3.1.&EXT";
   plot mn*time = trtnum/vaxis = axis2 haxis = axis1 hminor
= 1 vminor = 1 noframe caxis = STGB href = 1 href = 3.5
href = 6.5 href = 9.5 des = "Figure 3.1 A Simple Line Plot";
   format time timedf. trtnum trtdfb.;
   label trtnum = 'Legend:';
   run;

   FILENAME GSASFILE "&OUTLOC./Figure 3.2.&EXT";
   plot mn*time = trtnum/vaxis = axis2 haxis = axis1 hminor
= 1 vminor = 1 noframe caxis = STGB href = 1 href = 3.5
href = 6.5 href = 9.5 anno = iop_ANNO des = "Figure 3.2
With the SD Bar Displayed";
   format time timedf. trtnum trtdfb.;
   label trtnum = 'Legend:';
   run;
quit;
```

TABLE 3.3

Annotated Dataset Used in SGPLOT (ANNO_LABEL)

function	x1	y1	x1space	y1space	anchor	label	textcolor
text	1	30	datavalue	datavalue	left	Baseline	blue
text	3.5	30	datavalue	datavalue	left	Week 4	blue
text	6.5	30	datavalue	datavalue	left	Week 8	blue
text	9.5	30	datavalue	datavalue	left	Week 12	blue

3.3.2.1.5 SG Annotation Dataset for Displaying the Reference Line Labels in SGPLOT

- The reference labels of "Baseline", "Week 4", "Week 8," and "Week 12" are placed in the positions corresponding to the (X,Y) coordinates of (1, 30), (3.5, 30), (6.5, 30), and (9.5, 30) by using the function = "text" (Table 3.3).

- The "x1space" and "y1space" variables are used to specify the drawing space of X and Y coordinates. They are similar to "xsys" and "ysys" used in an SAS/GRAPH annotated dataset. The "datavalue" is used for both coordinates to position label text with respect to the data values.

- The SG annotation facility is one of the two new features added in SAS 9.3 for ODS Graphics (SAS Institute Inc., 2012). Another new feature in SAS 9.3 is the attribute map, which is used to define the visual attributes of particular group values (Matange and Heath, 2011).

3.3.2.1.6 HighLow Statements for Drawing the SD Bars in SGPLOT

The *SERIES* statement in SGPLOT is used to generate the line plots and the *HighLow* statement is used to draw the SD bar at each hour. For the two treatments' mean values at each hour, the smaller value has the *LOW* value set as *"MEAN - SD"* and the HIGH value set as the mean value; and the higher mean value has the mean IOP itself set as the LOW value and *'MEAN + SD'* set as the *HIGH* value. This is done in the dataset level (MNIOP_HL).

```
proc sgplot data = iop_hl des = "- A Line Plot with SD
Bar Displayed" sganno = anno_label;
  series x = time y = drug_mn/lineattrs = (thickness = 2
PATTERN = solid COLOR = red) MARKERS MARKERATTRS =
(SYMBOL = square COLOR = red) name = "drug" legendlabel =
"New Drug (N = &N_Trt1.)";
  highlow x = time high = drug_high low = drug_low/
lineattrs = (thickness = 2 PATTERN = solid COLOR = red);

  series x = time y = cntl_mn/lineattrs = (thickness = 2
PATTERN = shortdash COLOR = black) MARKERS MARKERATTRS =
```

```
(SYMBOL = CIRCLE COLOR = black) name = "cntl" legendlabel
= "Active Control (N = &N_Trt2.)";
  highlow x = time high = cntl_high low = cntl_low/
lineattrs = (thickness = 2 PATTERN = shortdash COLOR =
black);

  xaxis VALUES = (1 to 12 by 1) label = "Hours at Each
Visit";
  yaxis VALUES = (10 to 30 by 1) label = "Mean Average
Eye IOP (mm Hg)";
  REFLINE 0.5 3.5 6.5 9.5/AXIS = x LABELPOS = MIN
labelloc = inside;
  format time timedf.;
  keylegend "drug" "cntl"/noborder title = 'Legend:';
run;
```

3.3.2.2 Main Features of the Second SAS Program

3.3.2.2.1 Data Manipulation

- Proc Means is used to calculate and save the mean values of mortgage rates and points to intermediate datasets (rate_sum and point_sum).
- The intermediate datasets (rate_sum and point_sum) are further manipulated to prepare datasets ready for plotting in GPLOT and SGPLOT (ratepoint_gplot and ratepoint_sgplot).

3.3.2.2.2 Producing the Two Mortgage Line Plots in GPLOT

- The interpolation option *"INTERPOL = HILOJ"* in the symbol statement is used to display a line connecting the maximum (Hi) and minimum (Lo) values of rates and points for each year. To use this "HILO" feature, the rate and point for each year need to be presented in the order of average, maximum, and minimum values. Please note the data structure of "ratepoint_gplot" is different than that of "ratepoint_sgplot".
- The "LINE" option is used to display a solid line (with the value of 1) for mortgage rates and a dashed line for mortgage points (with the value of 2). Both rates and points are displayed in the same figure with the same y-axis using the "overlay" plot statement in Figure 3.3.

```
symbol1 interpol = hiloj cv = black ci = black w = 2 Line = 1;
symbol2 interpol = hiloj cv = red ci = red w = 2 Line = 2;
```

- The PLOT2 statement is used to put the mortgage point on the second y-axis with a new AXIS statement (AXIS3) in Figure 3.4.
- RUN-group is used to produce the two mortgage line plots within the same Proc GPLOT with upper and lower level bookmarks (a good feature to use to produce figures in PDF file format).

```
FILENAME GSASFILE "&OUTLOC./Figure 3.3.&EXT";
ods proclabel = "Rates and Points for 30 Year Fixed-Rate
Mortgage";
ods listing;
proc gplot data = ratepoint_gplot;
  plot rate*year point*year/overlay haxis = axis1 vaxis =
axis2 legend = legend1 href = (1972 to 2012 by 5) LHREF =
2 des = "- Figure 3.3 Rates and Points in the Same Y-axis";
  label rate = "Rate" point = "Point";
  run;

FILENAME GSASFILE "&OUTLOC./Figure 3.4.&EXT";
  plot rate*year/haxis = axis1 vaxis = axis2 legend =
legend1 href = (1972 to 2012 by 5) LHREF = 2 des =
"- Figure 3.4 Rates and Points in Different Y-axis";
  plot2 point*year/vaxis = axis3 legend = legend2;
  label rate = "Rate" point = "Point";
  run;
quit;
```

3.3.2.2.3 Producing the Two Mortgage Line Plots in SGPLOT

- A SERIES plot statement is used to draw line plots for the average mortgage rates and points from 1972 to 2012.
- A HighLow plot statement is used to display the maximum and minimum mortgage rate and point values for each year. In producing Figure 3.3, which has both the mortgage rates and points displayed on the same y-axis, one YAXIS statement is used to specify the data range and label for both rates and points. A KEYLEGEND statement is used to show legends to distinguish the rate and point lines.

```
proc sgplot data = ratepoint_sgplot des = "- Figure 3.3
Rates and Points in the Same Y-axis";
  series x = year y = rate_avg/lineattrs = (thickness = 2
PATTERN = solid COLOR = black)MARKERS MARKERATTRS =
(SYMBOL = circlefilled COLOR = black) name = "rate"
legendlabel = "Rate";
  highlow x = year high = rate_max low = rate_min/
lineattrs = (thickness = 2 PATTERN = solid COLOR = black);
  series x = year y = point_avg/lineattrs = (thickness = 2
PATTERN = shortdash COLOR = red) MARKERS MARKERATTRS =
(SYMBOL = circle COLOR = red) name = "point" legendlabel
= "Point";
  highlow x = year high = point_max low = point_min/
lineattrs = (thickness = 2 PATTERN = shortdash COLOR = red);

  xaxis VALUES = (1972 to 2012 by 5) label = "Year";
```

```
   yaxis VALUES = (0 to 20 by 1) label = "Rate and Point
(%)";
   refline (1972 to 2012 by 5)/axis = x lineattrs =
(pattern = shortdash);
   keylegend "rate" "point"/noborder title = 'Legend:';
run;
```

- In Figure 3.4, rates and points are displayed in different y-axes by using "YAXIS" for the mortgage rate and "Y2AXIS" for the point. The auto-legend is suppressed using the "NOAUTOLEGEND" option, and the labels for mortgage rates and points are directly placed at the end of each line using the "CURVELABEL" option. Using direct labeling to place identifying labels close to the object itself is a good way to make good graphs (Matange, 2013).

```
proc sgplot data = ratepoint_sgplot noautolegend des =
"- Figure 3.4 Rates and Points in Different Y-axes";
   series x = year y = rate_avg/lineattrs = (thickness = 2
PATTERN = solid COLOR = black) MARKERS MARKERATTRS =
(SYMBOL = circlefilled COLOR = black) curvelabel = "Rate";
   highlow x = year high = rate_max low = rate_min/
lineattrs = (thickness = 2 PATTERN = solid COLOR = black);

series x = year y = point_avg/
   lineattrs = (thickness = 2 PATTERN = shortdash COLOR =
red) y2axis MARKERS MARKERATTRS = (SYMBOL = circle COLOR
= red) curvelabel = "Point";
   highlow x = year high = point_max low = point_min/
lineattrs = (thickness = 2 PATTERN = shortdash COLOR = red)
y2axis;

   xaxis VALUES = (1972 to 2012 by 5) label = "Year";
   yaxis VALUES = (0 to 20 by 1) label = "Rate (%)";
   y2axis VALUES = (0 to 3 by.5) label = "Point (%)";
   refline (1972 to 2012 by 5)/axis = x lineattrs =
(pattern = shortdash);
run;
```

3.4 Summary and Discussion

Line plots can be produced using both the GPLOT and SGPLOT procedures. Depending on the custom design or requirement of the line plots, one procedure might be easier to use than the other. However, the statements and settings used to produce line plots in GPLOT and SGPLOT are quite different. Table 3.4 summarizes the main features used to produce the line plots using the two procedures.

TABLE 3.4

Comparing PROC GPLOT and PROC SGPLOT in Producing Line Plots

Features	GPLOT in SAS/GRAPH	SGPLOT in ODS Graphics
Line plot	Using INTERPOL = join in SYMBOL statement	Using SERIES plot statement series x = time y = drug_mn
Line and marker attributes	Using SYMBOL statement	Using LINEATTRS and MARKERATTRS plot statements
Axis attributes	AXIS global statements and VAXIS/HAXIS plot statement options	XAXIS and YAXIS plot statements
Legend	Global LEGEND statement	LEGENDLABEL in MARKERS plot statement
Reference lines	HREF and VREF option in plot statement. Note: Reference labels are put in the right side of the reference lines by using j = r (justify = right) option in IOP line plots. Note: Does not support transparency.	AXIS = option in REFLINE plot statement: by default the axis is set to y. Note: Reference label is put in the middle position of the reference line automatically; the justify option is not available. Annotated facility is used to display the reference label in the preferred position and color in IOP line plots. Note: The transparency feature results in bitmap format figures even if the figure is saved in PS or EMF format (SAS 9.3.1).
SD Bar in IOP Line Plots	Annotate facility.	HighLow plot statement
Max and Min Bars in Mortgage Line Plots	SYMBOL statement: interpol = hiloj	HighLow plot statement
Two Y-axes	Two PLOT (PLOT, PLOT2) statements with two VAXIS (vaxis = axis2, vaxis = axis3)	Two SERIES statements with y2axis in the 2nd SERIES statement
RUN-group	Supports RUN-group. Can set up the same upper-level and different lower-level bookmarks for the two figures.	Does not support RUN-group; can only produce the same upper-level bookmarks for different figures.
Pros	Easy to save the individual figures in different formats (EPS, PS, EMF, etc.) Flexible in placing reference line labels Supports RUN-group	Easy to produce SD, Min and Max bars using the HighLow statement.
Cons	More difficult to produce the SD bar: Annotate is used	EPS format figure is not supported. Not flexible with reference line labels: SG Annotate is used

The sample figures are produced and saved in the postscript (PS) listing format by using the PSCOLOR device driver in SGPLOT and by specifying OUTPUTFMT = PS in SGPLOT. Using the ODS PDF statement, we can produce the PDF document output file. Other listing format figures (EMF, etc.) can be produced by using the corresponding device driver (SASEMF, etc.) in GPLOT and output format in SGPLOT. RTF document files can be produced using ODS RTF together with the EMF format figures.

The bitmap format figures (e.g., PNG, JPEG, etc.) can be produced by adding XMAX, YMAX, XPIXELS, and YPIXELS options on the GOPTIONS statement in GPLOT, or by adding the statement "ods listing image_dpi = DpiValue" in SGPLOT.

3.5 References

Freddie Mac. 2013. "Historical PMMS Data," http://www.freddiemac.com/pmms/.

Matange, S., and Heath, D. 2011. *Statistical Graphics Procedures by Example: Effective Graphs Using SAS®*. Cary, NC: SAS Institute Inc.

Matange, S. 2013. "Make a Good Graph." In *SAS Global Forum 2013 Proceedings*. Cary, NC: SAS Institute Inc., http://support.sas.com/resources/papers/proceedings13/361-2013.pdf.

SAS Institute Inc. 2012. *SAS/GRAPH® 9.3: Reference*, 3rd ed. Cary, NC: SAS Institute Inc.

SAS Institute Inc. 2012. "SG Annotation." In *SAS® 9.3 ODS Graphics: Procedures Guide*, 3rd ed. Cary, NC: SAS Institute Inc.

3.6 Appendix: SAS Programs for Producing the Sample Figures

SAS programs "Line Plots for IOP Values.sas" and "Line Plots for Mortgage Rates and Points by Year.sas" were used to produce the four sample figures included.

3.6.1 Line Plots for IOP Values

```
******************************************************************;
* Program Name: Line Plots for IOP Values.sas                   *;
* Function: Produce the following two figures in both GPLOT     *;
* and SGPLOT                                                    *;
* -. Figure 3.1 Mean Average Eye IOP at Each Visit and Hour    *;
* -. Figure 3.2 Mean Average Eye IOP at Each Visit and Hour    *;
* with SD Displayed                                            *;
******************************************************************;
```

```
options mprint symbolgen nodate nonumber validvarname = v7
orientation = landscape;
%let pgmname = Line Plots for IOP Values.sas;
%let pgmloc = C:\SASBook\SAS Programs;
%let pgmpth = &pgmloc.\&pgmname. &sysdate9. &systime. SAS
V&sysver.;
%let outloc = C:\SASBook\Sample Figures\Chapter 3;

** Set-up the site, subject number and SD for data simulation;
%let sitenum = 10;
%let subjnum = 500;
%let SD = 3.5;

proc format;
   value trtdf
     1 = 'New Drug'
     2 = 'Active Control'
     OTHER = ' ';
   value visdf
     1 = 'Baseline'
     2 = 'Week 4'
     3 = 'Week 8'
     4 = 'Week 12';
   value hrdf
        1 = 'Hour 0'
        2 = 'Hour 2'
        3 = 'Hour 8';
run;

** Generate the required number of subjects;
data subj;
   do i = 1 to &subjnum.;
     subjid = 1000+ i;
     shuffle = ranuni (0);
     output;
   end;
   drop i;
run;

proc sort data = subj;
   by shuffle;
run;

** Randomly assign the subjects to 8 sites;
** Within each site randomly assign 2 treatment groups;
data site_subj;
   set subj;
   if shuffle <.1 then do;
     site = 1;
     if ranuni (0) < 0.5 then trtnum = 1;
```

```
        else trtnum = 2;
    end;
    else if.1 < = shuffle <.30 then do;
      site = 2;
      if ranuni (0) < 0.5 then trtnum = 1;
        else trtnum = 2;
    end;
    else if.30 < = shuffle <.35 then do;
      site = 3;
      if ranuni (0) < 0.5 then trtnum = 1;
        else trtnum = 2;
    end;
    else if.35 < = shuffle <.5 then do;
      site = 4;
      if ranuni (0) < 0.5 then trtnum = 1;
        else trtnum = 2;
    end;
    else if.5 < = shuffle <.6 then do;
      site = 5;
      if ranuni (0) < 0.5 then trtnum = 1;
        else trtnum = 2;
    end;
    else if.6 < = shuffle <.8 then do;
      site = 6;
      if ranuni (0) < 0.5 then trtnum = 1;
        else trtnum = 2;
    end;
    else if.8 < = shuffle <.85 then do;
      site = 7;
      if ranuni (0) < 0.5 then trtnum = 1;
        else trtnum = 2;
    end;
    else if.85 < = shuffle < = 1.0 then do;
      site = 8;
      if ranuni (0) < 0.5 then trtnum = 1;
        else trtnum = 2;
    end;
run;

proc freq data = site_subj noprint;
  table trtnum/out = subj_trt;
run;

** Save the subject number at each treatment group to macro
variables for later use;
data _null_;
  set subj_trt;
  if trtnum = 1 then call symput ("N_Trt1", put(count, 3.0));
  if trtnum = 2 then call symput ("N_Trt2", put(count, 3.0));
run;
```

```
** Set up the IOP Values based on the trt assignment and
visits/timepoints;
data iop;
  set site_subj;
  do i = 1 to 4; ** 4 visits;
    do j = 1 to 3; ** 3 timepoints/visit;
      visit = i;
      hour = j;
      if i = 1 then do; ** Baseline;
        if j = 1 then iop = round((RANNOR(0)* &SD. + 25),.1);
** Hour 0;
        if j = 2 then iop = round((RANNOR(0)* &SD. + 23),.1);
** Hour 2;
        if j = 3 then iop = round((RANNOR(0)* &SD. + 22),.1);
** Hour 8;
      end;
      else if i > 1 and trtnum = 1 then do; ** Post-baseline:
New Drug;
        if j = 1 then iop = round((RANNOR(0)* &SD.
+ 17.5),.1); ** Hour 0;
        if j = 2 then iop = round((RANNOR(0)* &SD.
+ 16.5),.1); ** Hour 2;
        if j = 3 then iop = round((RANNOR(0)* &SD.
+ 16.2),.1); ** Hour 8;
      end;
      else if i > 1 and trtnum = 2 then do; ** Post-baseline:
Active Control;
        if j = 1 then iop = round((RANNOR(0)* &SD. + 20),.1);
** Hour 0;
        if j = 2 then iop = round((RANNOR(0)* &SD. + 19),.1);
** Hour 2;
        if j = 3 then iop = round((RANNOR(0)* &SD. + 18.7),.1);
** Hour 8;
      end;
      output;
    end;
  end;
  drop i j shuffle;
  format visit visdf. hour hrdf. trtnum trtdf.;
run;

proc sort data = iop;
  by site subjid visit hour;
run;

proc means data = iop noprint;
  class visit hour trtnum;
  var iop;
  output out = iop_stat n = n mean = mn std = sd;
run;
```

```
data MN_IOP;
  set iop_stat;
  where visit ne. and hour ne. and trtnum ne.;
  if visit = 1 then do; * Baseline;
    if hour = 1 then time = 1;
    if hour = 2 then time = 2;
    if hour = 3 then time = 3;
  end;
  if visit = 2 then do; * Week 4;
    if hour = 1 then time = 4;
    if hour = 2 then time = 5;
    if hour = 3 then time = 6;
  end;
  if visit = 3 then do; * Week 8;
    if hour = 1 then time = 7;
    if hour = 2 then time = 8;
    if hour = 3 then time = 9;
  end;
  if visit = 4 then do; * Week 12;
    if hour = 1 then time = 10;
    if hour = 2 then time = 11;
    if hour = 3 then time = 12;
  end;
  drop _TYPE_ _FREQ_;
run;

** To display SD bars using the annotate facility;
** The SD bar is displayed in the upper position for the trt;
** with the larger mean and in the lower position for the trt;
** with the smaller mean at the same visit/hour;
proc sort data = mn_iop;
  by time mn;
run;

** Make the SAS Annotate data set macros available for use;
%ANNOMAC;

data iop_ANNO;
  set mn_iop;
  %DCLANNO;
  RETAIN SDMULT 1 NUM_OFFSET.1 SHIFT_VAL.15;
  SIZE = 2;
  HSYS = '4';
  XSYS = '2';
  YSYS = '2';

  UPPER_LIM = mn + (SDMULT*sd); /* Y at top of SD */
  LOWER_LIM = mn - (SDMULT*sd); /* Y at bottom of SD */
```

```
  if Mod(_n_, 2) = 1 then do; ** odd number: with the smaller
mean;
    if trtnum = 1 then do;
      %LINE(time, mn, time, LOWER_LIM, red, 1, SIZE);
    end;
    if trtnum = 2 then do;
      %LINE(time, mn, time, LOWER_LIM, black, 2, SIZE);
    end;
  end;
  if Mod(_n_, 2) = 0 then do; ** even number: with the higher
mean;
    if trtnum = 1 then do;
      %LINE(time, mn, time, UPPER_LIM, red, 1, SIZE);
    end;
    if trtnum = 2 then do;
      %LINE(time, mn, time, UPPER_LIM, black, 2, SIZE);
    end;
  end;
  KEEP X Y FUNCTION COLOR LINE SIZE HSYS XSYS YSYS STYLE
LOWER_LIM UPPER_LIM time trtnum mn sd;
RUN;

proc format;
  value timedf
    1, 4, 7, 10 = '8 AM'
    2, 5, 8, 11 = '10 AM'
    3, 6, 9, 12 = '4 PM'
    OTHER = ' ';
  value trtdfb
    1 = "New Drug (N = &N_Trt1.) "
    2 = "Active Control (N = &N_Trt2.)"
  OTHER = ' ';
run;

%LET FONTNAME = Times;%LET DRIVER = PSCOLOR;%LEt EXT = PS;
goptions
  reset    = all
  GUNIT    = PCT
  rotate   = landscape
  gsfmode  = replace
  gsfname  = GSASFILE
  device   = &DRIVER
  lfactor  = 2
  hsize    = 8 in
  horigin  = 0 in
  vsize    = 6 in
  vorigin  = 0 in
  ftext    = "&FONTNAME"
  htext    = 10pt
  ftitle   = "&FONTNAME"
```

```
   htitle   = 10pt
;

data iop_ANNO;
   set iop_ANNO;
   style = "'&fontname.'";
run;

SYMBOL1 H = 2 C = RED      CO = RED INTERPOL = join W = 2 L = 1
VALUE = SQUARE;
SYMBOL2 H = 3 C = Black     CO = Black INTERPOL = join W = 2
L = 2 VALUE = CIRCLE;
axis1 minor = none order = (1 to 12 by 1)
   label = (h = 2.5 font = "&FONTNAME" "Hours at Each Visit")
   reflabel = (position = top c = blue font = "&FONTNAME" h = 2
j = r "Baseline" "Week 4" "Week 8" "Week 12");
axis2 minor = none order = (10 to 30 by 1)
   label = (a = 90 r = 0 h = 2.5 font = "&fontname" "Mean IOP
(mm Hg)");
ods proclabel = "Mean IOP by Treatment at Each Visit/Hour";
proc gplot data = MN_IOP;
   title1 "Figure 3.1 Mean IOP by Treatment at Each Visit/Hour";
   title2 "- A Simple Line Plot Produced in GPLOT";
   footnote1 "&pgmpth.";
   FILENAME GSASFILE "&OUTLOC./Figure 3.1.&EXT";
   plot mn*time = trtnum/vaxis = axis2 haxis = axis1 hminor = 1
vminor = 1 noframe caxis = STGB href = 1 href = 3.5 href = 6.5
href = 9.5 des = "Figure 3.1 A Simple Line Plot";
      format time timedf. trtnum trtdfb.;
      label trtnum = 'Legend:';
   run;

   title1 "Figure 3.2 Mean IOP by Treatment at Each Visit/Hour";
   title2 "-  A Line Plot with SD Bar Produced in GPLOT";
   footnote1 "&pgmpth.";
   FILENAME GSASFILE "&OUTLOC./Figure 3.2.&EXT";
   plot mn*time = trtnum/vaxis = axis2 haxis = axis1 hminor = 1
vminor = 1 noframe caxis = STGB
      href = 1 href = 3.5 href = 6.5 href = 9.5 anno = iop_ANNO
      des = "Figure 3.2 With the SD Bar Displayed";
      format time timedf. trtnum trtdfb.;
      label trtnum = 'Legend:';
   run;
quit;

***************************************************************;
** Reproduce the line plots using the SGPLOT procedure     *;
***************************************************************;
proc transpose data = mn_iop out = mniop (rename = (new_drug =
drug_mn active_control = cntl_mn));
```

```
   by time;
   var mn;
   id trtnum;
run;

proc transpose data = mn_iop out = sdiop (rename = (new_drug =
drug_sd active_control = cntl_sd));
   by time;
   var sd;
   id trtnum;
run;

data iop_hl;
   merge mniop sdiop;
   by time;
   if drug_mn > = cntl_mn then do;
     drug_high = drug_mn + drug_sd; drug_low = drug_mn;
     cntl_high = cntl_mn; cntl_low = cntl_mn - cntl_sd;
   end;
   else if drug_mn < cntl_mn then do;
     drug_high = drug_mn; drug_low = drug_mn - drug_sd;
     cntl_high = cntl_mn + cntl_sd; cntl_low = cntl_mn;
   end;
run;

data anno_label;
   function = "text"; x1 = 1; y1 = 30; x1space = "datavalue";
   y1space = "datavalue"; anchor = "left";
   label = "Baseline"; textcolor = "blue"; output;

   function = "text"; x1 = 3.5; y1 = 30; x1space = "datavalue";
   y1space = "datavalue"; anchor = "left";
   label = "Week 4"; textcolor = "blue"; output;

   function = "text"; x1 = 6.5; y1 = 30; x1space = "datavalue";
   y1space = "datavalue"; anchor = "left";
   label = "Week 8"; textcolor = "blue"; output;

   function = "text"; x1 = 9.5; y1 = 30; x1space = "datavalue";
   y1space = "datavalue"; anchor = "left";
   label = "Week 12"; textcolor = "blue"; output;
run;

%LET OUTPUTFMT = PS;
ods listing gpath = "&outloc.";
ods graphics/reset = all width = 8in height = 6in noborder
OUTPUTFMT = &OUTPUTFMT. imagename = "FigSG 3_1";
ods proclabel = "Mean IOP by Treatment at Each Visit/Hour";
title1 "Figure 3.1 Mean IOP by Treatment at Each Visit/Hour";
title2 "- A Simple Line Plot Produced in SSGPLOT";
```

```
footnote1 "&pgmpth.";
proc sgplot data = iop_hl des = "- A Simple Line Plot" sganno
= anno_label;
  series x = time y = drug_mn/lineattrs = (thickness = 2
PATTERN = solid COLOR = red) MARKERS MARKERATTRS = (SYMBOL =
square COLOR = red) name = "drug" legendlabel = "New Drug
(N = &N_Trt1.)";
  series x = time y = cntl_mn/lineattrs = (thickness = 2
PATTERN = shortdash COLOR = black) MARKERS MARKERATTRS =
(SYMBOL = CIRCLE COLOR = black) name = "cntl" legendlabel =
"Active Control (N = &N_Trt2.)";

  xaxis VALUES = (1 to 12 by 1) label = "Hours at Each Visit";
  yaxis VALUES = (10 to 30 by 1) label = "Mean IOP (mm Hg)";
  REFLINE 0.5 3.5 6.5 9.5/AXIS = x;
  format time timedf.;
  keylegend "drug" "cntl"/noborder title = 'Legend:';
run;
quit;

ods listing gpath = "&outloc.";
ods graphics/reset = all width = 8in height = 6in noborder
OUTPUTFMT = &OUTPUTFMT. imagename = "FigSG 3_2";
ods proclabel = "Mean IOP by Treatment at Each Visit/Hour";
title1 "Figure 3.2 Mean IOP by Treatment at Each Visit/Hour";
title2 "- A Line Plot with SD Bar Produced in SGPLOT";
footnote1 "&pgmpth.";
proc sgplot data = iop_hl des = "- A Line Plot with SD Bar
Displayed" sganno = anno_label;
  series x = time y = drug_mn/lineattrs = (thickness = 2
PATTERN = solid COLOR = red) MARKERS MARKERATTRS = (SYMBOL =
square COLOR = red) name = "drug" legendlabel = "New Drug
(N = &N_Trt1.)";
  highlow x = time high = drug_high low = drug_low/lineattrs =
(thickness = 2 PATTERN = solid COLOR = red);
  series x = time y = cntl_mn/lineattrs = (thickness = 2
PATTERN = shortdash COLOR = black) MARKERS MARKERATTRS =
(SYMBOL = CIRCLE COLOR = black) name = "cntl" legendlabel =
"Active Control (N = &N_Trt2.)";
  highlow x = time high = cntl_high low = cntl_low/lineattrs =
(thickness = 2 PATTERN = shortdash COLOR = black);

  xaxis VALUES = (1 to 12 by 1) label = "Hours at Each Visit";
  yaxis VALUES = (10 to 30 by 1) label = "Mean IOP (mm Hg)";
  REFLINE 0.5 3.5 6.5 9.5/AXIS = x LABELPOS = MIN labelloc =
inside;
  format time timedf.;
  keylegend "drug" "cntl"/noborder title = 'Legend:';
run;
quit;
```

3.6.2 Line Plots for Mortgage Rates and Points

```
****************************************************************;
* Program Name: Line Plots for Mortgage Rates and Points by *;
* Year.sas                                                  *;
* Function: Produce the following two figures in both GPLOT *;
* and SGPLOT                                                *;
* -. Figure 5.3 Rates and Points for 30 Year Fixed-Rate     *;
* Mortgage                                                  *;
* - 1972 to 2012: with Rates and Points Displayed in the    *;
* Same Y-axis                                               *;
* -. Figure 5.3 Rates and Points for 30 Year Fixed-Rate     *;
* Mortgage                                                  *;
* - 1972 to 2012: with Rates and Points Displayed in        *;
* Different Y-axis                                          *;
****************************************************************;
options mprint symbolgen nodate nonumber validvarname = v7
orientation = landscape;
%let pgmname = Line Plots for Mortgage Rates and Points by
Year.sas;
%let pgmloc = C:\SASBook\SAS Programs;
%let outloc = C:\SASBook\Sample Figures\Chapter 3;
%let dataloc = C:\SASBook\Data;
%let pgmpth = &pgmloc.\&pgmname. &sysdate9. &systime. SAS
V&sysver.;

data FixedRate;
  input Year Month Rate Point @@;
  datalines;
1972 1 7.44 1.0 1972 2 7.32 0.9 1972 3 7.29 0.9 1972 4 7.29 0.9
1972 5 7.37 0.9 1972 6 7.37 0.9 1972 7 7.40 0.9 1972 8 7.40 0.9
1972 9 7.42 1.0 1972 10 7.42 1.0 1972 11 7.43 1.0 1972 12 7.44 1.0
1973 1 7.44 0.9 1973 2 7.44 1.0 1973 3 7.46 0.9 1973 4 7.54 0.9
1973 5 7.65 0.9 1973 6 7.73 0.9 1973 7 8.05 1.0 1973 8 8.50 1.0
1973 9 8.82 1.1 1973 10 8.77 1.1 1973 11 8.58 1.0 1973 12 8.54 1.0
1974 1 8.54 1.0 1974 2 8.46 1.0 1974 3 8.41 1.0 1974 4 8.58 1.0
1974 5 8.97 1.1 1974 6 9.09 1.2 1974 7 9.28 1.3 1974 8 9.59 1.3
1974 9 9.96 1.4 1974 10 9.98 1.5 1974 11 9.79 1.4 1974 12 9.62 1.3
1975 1 9.43 1.2 1975 2 9.10 1.2 1975 3 8.89 1.1 1975 4 8.82 1.0
1975 5 8.91 1.1 1975 6 8.89 1.0 1975 7 8.89 1.1 1975 8 8.94 1.1
1975 9 9.12 1.1 1975 10 9.22 1.1 1975 11 9.15 1.1 1975 12 9.10 1.1
1976 1 9.02 1.1 1976 2 8.81 1.0 1976 3 8.76 1.3 1976 4 8.73 1.3
1976 5 8.76 1.3 1976 6 8.85 1.3 1976 7 8.93 1.2 1976 8 9.00 1.2
1976 9 8.98 1.2 1976 10 8.92 1.2 1976 11 8.81 1.3 1976 12 8.79 1.2
1977 1 8.72 1.1 1977 2 8.67 1.1 1977 3 8.69 1.2 1977 4 8.75 1.1
1977 5 8.83 1.1 1977 6 8.86 1.1 1977 7 8.94 1.1 1977 8 8.94 1.1
1977 9 8.90 1.1 1977 10 8.92 1.2 1977 11 8.92 1.1 1977 12 8.96 1.2
1978 1 9.01 1.3 1978 2 9.14 1.3 1978 3 9.20 1.3 1978 4 9.35 1.3
1978 5 9.57 1.3 1978 6 9.71 1.4 1978 7 9.74 1.4 1978 8 9.78 1.3
1978 9 9.76 1.3 1978 10 9.86 1.2 1978 11 10.11 1.2 1978 12 10.35 1.4
1979 1 10.39 1.5 1979 2 10.41 1.5 1979 3 10.43 1.5 1979 4 10.50 1.5
```

```
1979 5 10.69 1.6 1979 6 11.04 1.6 1979 7 11.09 1.7 1979 8 11.09 1.7
1979 9 11.30 1.6 1979 10 11.64 1.7 1979 11 12.83 1.7 1979 12 12.90 1.6
1980 1 12.88 1.6 1980 2 13.04 1.6 1980 3 15.28 2.0 1980 4 16.32 1.9
1980 5 14.26 1.9 1980 6 12.71 1.8 1980 7 12.19 1.8 1980 8 12.56 1.7
1980 9 13.20 1.7 1980 10 13.79 1.7 1980 11 14.21 1.7 1980 12 14.79 1.7
1981 1 14.90 2.0 1981 2 15.13 2.0 1981 3 15.40 2.0 1981 4 15.58 2.0
1981 5 16.40 2.1 1981 6 16.70 2.1 1981 7 16.83 2.1 1981 8 17.28 2.1
1981 9 18.16 2.1 1981 10 18.45 2.3 1981 11 17.82 2.1 1981 12 16.95 2.1
1982 1 17.48 2.2 1982 2 17.60 2.2 1982 3 17.16 2.2 1982 4 16.89 2.3
1982 5 16.68 2.3 1982 6 16.70 2.2 1982 7 16.82 2.2 1982 8 16.27 2.3
1982 9 15.43 2.3 1982 10 14.61 2.2 1982 11 13.82 2.2 1982 12 13.62 2.2
1983 1 13.25 2.2 1983 2 13.04 2.0 1983 3 12.80 2.2 1983 4 12.78 2.1
1983 5 12.63 2.1 1983 6 12.87 2.1 1983 7 13.43 2.2 1983 8 13.81 2.2
1983 9 13.73 2.2 1983 10 13.54 2.1 1983 11 13.44 2.1 1983 12 13.42 2.2
1984 1 13.37 2.3 1984 2 13.23 2.4 1984 3 13.39 2.4 1984 4 13.65 2.4
1984 5 13.94 2.5 1984 6 14.42 2.5 1984 7 14.67 2.6 1984 8 14.47 2.6
1984 9 14.35 2.6 1984 10 14.13 2.6 1984 11 13.64 2.5 1984 12 13.18 2.5
1985 1 13.08 2.5 1985 2 12.92 2.4 1985 3 13.17 2.6 1985 4 13.20 2.6
1985 5 12.91 2.5 1985 6 12.22 2.5 1985 7 12.03 2.5 1985 8 12.19 2.6
1985 9 12.19 2.6 1985 10 12.14 2.5 1985 11 11.78 2.4 1985 12 11.26 2.3
1986 1 10.89 2.3 1986 2 10.71 2.3 1986 3 10.08 2.3 1986 4 9.94 2.2
1986 5 10.15 2.3 1986 6 10.69 2.3 1986 7 10.51 2.2 1986 8 10.20 2.1
1986 9 10.01 2.2 1986 10 9.98 2.1 1986 11 9.70 2.0 1986 12 9.32 2.1
1987 1 9.20 2.2 1987 2 9.08 2.1 1987 3 9.04 2.1 1987 4 9.83 2.3
1987 5 10.60 2.3 1987 6 10.54 2.2 1987 7 10.28 2.2 1987 8 10.33 2.1
1987 9 10.89 2.2 1987 10 11.26 2.2 1987 11 10.65 2.1 1987 12 10.64 2.1
1988 1 10.38 2.0 1988 2 9.89 2.1 1988 3 9.93 2.0 1988 4 10.20 2.1
1988 5 10.46 2.1 1988 6 10.46 2.0 1988 7 10.43 2.0 1988 8 10.60 2.2
1988 9 10.48 2.1 1988 10 10.30 1.9 1988 11 10.27 2.1 1988 12 10.61 2.1
1989 1 10.73 2.1 1989 2 10.65 2.2 1989 3 11.03 2.2 1989 4 11.05 2.2
1989 5 10.77 2.1 1989 6 10.20 2.1 1989 7 9.88 2.1 1989 8 9.99 2.1
1989 9 10.13 2.0 1989 10 9.95 2.0 1989 11 9.77 2.0 1989 12 9.74 2.0
1990 1 9.90 2.1 1990 2 10.20 2.1 1990 3 10.27 2.1 1990 4 10.37 2.1
1990 5 10.48 2.0 1990 6 10.16 2.0 1990 7 10.04 2.0 1990 8 10.10 2.0
1990 9 10.18 2.1 1990 10 10.17 2.2 1990 11 10.01 2.1 1990 12 9.67 1.9
1991 1 9.64 2.1 1991 2 9.37 2.0 1991 3 9.50 2.1 1991 4 9.50 2.0
1991 5 9.47 2.0 1991 6 9.62 2.1 1991 7 9.58 2.0 1991 8 9.24 1.9
1991 9 9.01 1.9 1991 10 8.86 1.9 1991 11 8.71 1.8 1991 12 8.50 1.8
1992 1 8.43 1.8 1992 2 8.76 1.8 1992 3 8.94 1.9 1992 4 8.85 1.7
1992 5 8.67 1.7 1992 6 8.51 1.7 1992 7 8.13 1.6 1992 8 7.98 1.7
1992 9 7.92 1.7 1992 10 8.09 1.8 1992 11 8.31 1.9 1992 12 8.21 1.6
1993 1 7.99 1.6 1993 2 7.68 1.5 1993 3 7.50 1.6 1993 4 7.46 1.7
1993 5 7.47 1.8 1993 6 7.42 1.6 1993 7 7.21 1.6 1993 8 7.11 1.5
1993 9 6.91 1.5 1993 10 6.83 1.5 1993 11 7.16 1.6 1993 12 7.17 1.7
1994 1 7.07 1.7 1994 2 7.15 1.8 1994 3 7.68 1.7 1994 4 8.32 1.8
1994 5 8.60 1.8 1994 6 8.40 1.8 1994 7 8.61 1.8 1994 8 8.51 1.8
1994 9 8.64 1.8 1994 10 8.93 1.8 1994 11 9.17 1.8 1994 12 9.20 1.8
1995 1 9.15 1.8 1995 2 8.83 1.9 1995 3 8.46 1.8 1995 4 8.32 1.9
1995 5 7.96 1.8 1995 6 7.57 1.8 1995 7 7.61 1.8 1995 8 7.86 1.8
1995 9 7.64 1.8 1995 10 7.48 1.9 1995 11 7.38 1.8 1995 12 7.20 1.8
1996 1 7.03 1.8 1996 2 7.08 1.7 1996 3 7.62 1.8 1996 4 7.93 1.8
1996 5 8.07 1.7 1996 6 8.32 1.7 1996 7 8.25 1.8 1996 8 8.00 1.7
1996 9 8.23 1.7 1996 10 7.92 1.7 1996 11 7.62 1.8 1996 12 7.60 1.7
```

```
1997 1 7.82 1.8 1997 2 7.65 1.7 1997 3 7.90 1.8 1997 4 8.14 1.7
1997 5 7.94 1.7 1997 6 7.69 1.7 1997 7 7.50 1.8 1997 8 7.48 1.7
1997 9 7.43 1.7 1997 10 7.29 1.7 1997 11 7.21 1.7 1997 12 7.10 1.8
1998 1 6.99 1.4 1998 2 7.04 1.2 1998 3 7.13 1.2 1998 4 7.14 1.0
1998 5 7.14 1.1 1998 6 7.00 1.0 1998 7 6.95 1.1 1998 8 6.92 1.1
1998 9 6.72 1.0 1998 10 6.71 0.9 1998 11 6.87 0.9 1998 12 6.74 1.0
1999 1 6.79 0.9 1999 2 6.81 1.0 1999 3 7.04 0.9 1999 4 6.92 1.0
1999 5 7.15 1.0 1999 6 7.55 1.0 1999 7 7.63 1.0 1999 8 7.94 1.0
1999 9 7.82 1.0 1999 10 7.85 1.0 1999 11 7.74 1.0 1999 12 7.91 1.0
2000 1 8.21 1.0 2000 2 8.33 1.0 2000 3 8.24 1.0 2000 4 8.15 1.0
2000 5 8.52 1.0 2000 6 8.29 0.9 2000 7 8.15 0.9 2000 8 8.03 1.0
2000 9 7.91 1.0 2000 10 7.80 1.0 2000 11 7.75 0.9 2000 12 7.38 1.0
2001 1 7.03 0.9 2001 2 7.05 1.0 2001 3 6.95 0.9 2001 4 7.08 0.9
2001 5 7.15 1.0 2001 6 7.16 1.0 2001 7 7.13 0.9 2001 8 6.95 0.9
2001 9 6.82 0.9 2001 10 6.62 0.9 2001 11 6.66 0.8 2001 12 7.07 0.8
2002 1 7.00 0.8 2002 2 6.89 0.7 2002 3 7.01 0.7 2002 4 6.99 0.7
2002 5 6.81 0.7 2002 6 6.65 0.6 2002 7 6.49 0.6 2002 8 6.29 0.6
2002 9 6.09 0.6 2002 10 6.11 0.6 2002 11 6.07 0.6 2002 12 6.05 0.6
2003 1 5.92 0.6 2003 2 5.84 0.6 2003 3 5.75 0.6 2003 4 5.81 0.6
2003 5 5.48 0.6 2003 6 5.23 0.6 2003 7 5.63 0.5 2003 8 6.26 0.7
2003 9 6.15 0.6 2003 10 5.95 0.6 2003 11 5.93 0.6 2003 12 5.88 0.7
2004 1 5.71 0.7 2004 2 5.64 0.7 2004 3 5.45 0.7 2004 4 5.83 0.7
2004 5 6.27 0.7 2004 6 6.29 0.6 2004 7 6.06 0.6 2004 8 5.87 0.7
2004 9 5.75 0.7 2004 10 5.72 0.7 2004 11 5.73 0.6 2004 12 5.75 0.6
2006 1 6.15 0.5 2006 2 6.25 0.6 2006 3 6.32 0.6 2006 4 6.51 0.6
2006 5 6.60 0.5 2006 6 6.68 0.5 2006 7 6.76 0.5 2006 8 6.52 0.4
2006 9 6.40 0.5 2006 10 6.36 0.4 2006 11 6.24 0.5 2006 12 6.14 0.4
2007 1 6.22 0.4 2007 2 6.29 0.4 2007 3 6.16 0.4 2007 4 6.18 0.5
2007 5 6.26 0.4 2007 6 6.66 0.4 2007 7 6.70 0.4 2007 8 6.57 0.4
2007 9 6.38 0.5 2007 10 6.38 0.5 2007 11 6.21 0.4 2007 12 6.10 0.5
2008 1 5.76 0.4 2008 2 5.92 0.5 2008 3 5.97 0.5 2008 4 5.92 0.4
2008 5 6.04 0.5 2008 6 6.32 0.7 2008 7 6.43 0.6 2008 8 6.48 0.7
2008 9 6.04 0.7 2008 10 6.20 0.6 2008 11 6.09 0.7 2008 12 5.29 0.7
2009 1 5.05 0.7 2009 2 5.13 0.7 2009 3 5.00 0.7 2009 4 4.81 0.7
2009 5 4.86 0.7 2009 6 5.42 0.7 2009 7 5.22 0.7 2009 8 5.19 0.7
2009 9 5.06 0.7 2009 10 4.95 0.7 2009 11 4.88 0.7 2009 12 4.93 0.7
2010 1 5.03 0.7 2010 2 4.99 0.7 2010 3 4.97 0.7 2010 4 5.10 0.7
2010 5 4.89 0.7 2010 6 4.74 0.7 2010 7 4.56 0.7 2010 8 4.43 0.7
2010 9 4.35 0.7 2010 10 4.23 0.8 2010 11 4.30 0.8 2010 12 4.71 0.7
2011 1 4.76 0.8 2011 2 4.95 0.7 2011 3 4.84 0.7 2011 4 4.84 0.7
2011 5 4.64 0.7 2011 6 4.51 0.7 2011 7 4.55 0.7 2011 8 4.27 0.7
2011 9 4.11 0.7 2011 10 4.07 0.8 2011 11 3.99 0.7 2011 12 3.96 0.7
2012 1 3.92 0.8 2012 2 3.89 0.8 2012 3 3.95 0.8 2012 4 3.91 0.7
2012 5 3.80 0.8 2012 6 3.68 0.7 2012 7 3.55 0.7 2012 8 3.60 0.6
2012 9 3.47 0.6 2012 10 3.38 0.7 2012 11 3.35 0.7 2012 12 3.35 0.7
;
run;

** Min, Max and Average Rate by Year;
proc means data = FixedRate noprint;
  where month ne 13 and year ne 2013;
  by year;
```

```
   var rate;
   output out = rate_sum mean = rate_avg min = rate_min max =
rate_max;
run;

** Min, Max and Average Point by Year;
proc means data = FixedRate noprint;
   where month ne 13 and year ne 2013;
   by year;
   var point;
   output out = point_sum mean = point_avg min = point_min max
= point_max;
run;

data rate_sum2;
   set rate_sum;
   drop rate_avg rate_min rate_max _TYPE_ _FREQ_;
   rate = rate_avg; sum = 'Avg'; output;
   rate = rate_max; sum = 'Max'; output;
   rate = rate_min; sum = 'Min'; output;
run;

data point_sum2;
   set point_sum;
   drop point_avg point_min point_max _TYPE_ _FREQ_;
   point = point_avg; sum = 'Avg'; output;
   point = point_max; sum = 'Max'; output;
   point = point_min; sum = 'Min'; output;
run;

** data for gplot;
data ratepoint_gplot;
   merge rate_sum2 point_sum2;
   by year sum;
run;

** data for sgplot;
data ratepoint_sgplot;
   merge rate_sum point_sum;
   by year;
   drop _FREQ_ _TYPE_;
run;

%LET FONTNAME = Times;
%LET DRIVER = PSCOLOR;%LEt EXT = PS;
goptions
   reset    = all
   GUNIT    = PCT
   rotate   = landscape
   gsfmode  = replace
```

```
   gsfname = GSASFILE
   device  = &DRIVER.
   lfactor = 2
   hsize   = 8 in
   horigin = 0 in
   vsize   = 6 in
   vorigin = 0 in
   ftext   = "&FONTNAME."
   htext   = 10pt
   ftitle  = "&FONTNAME."
   htitle  = 10pt
;

symbol1 interpol = hiloj cv = black ci = black w = 2 Line = 1;
symbol2 interpol = hiloj cv = red ci = red w = 2 Line = 2;
legend1 label = ("Legend: ");
axis1 order = (1972 to 2012 by 5) major = (h = 1 w = 1) minor
= (number = 4 h =.5 w = 1) value = (h = 2) label = (h = 2.5
font = "&FONTNAME" "Year") offset = (3,3);
axis2 order = (0 to 20 by 1) minor = none value = (h = 2)
label = (a = 90 r = 0 h = 2.5 font = "&fontname" "30-Year
Fixed Mortgage Rate and Point(%)") offset = (2,2);

title1 "Figure 3.3 Rates and Points for 30 Year Fixed-Rate
Mortgage: Maximum-Average-Minimum";
title2 "1972 to 2012: with Rates and Points Displayed in the
Same Y-axis";
title3 "Produced Using GPLOT";
footnote1 "&pgmpth.";
FILENAME GSASFILE "&OUTLOC./Figure 3.3.&EXT";
ods proclabel = "Rates and Points for 30 Year Fixed-Rate
Mortgage";
ods listing;
proc gplot data = ratepoint_gplot;
   plot rate*year point*year/overlay haxis = axis1 vaxis = axis2
     legend = legend1 href = (1972 to 2012 by 5) LHREF = 2
     des = "- Figure 3.3 Rates and Points in the Same Y-axis";
   label rate = "Rate" point = "Point";
run;

axis2 order = (0 to 20 by 1) minor = none value = (h = 2)
   label = (a = 90 r = 0 h = 2.5 font = "&fontname" "30-Year
Fixed Mortgage Rate (%)") offset = (2,2);
axis3 order = (0 to 3 by 0.5) minor = none value = (h = 2)
   label = (a = 90 r = 0 h = 2.5 font = "&fontname" "30-Year
Fixed Mortgage Point (%)") offset = (2,2);
legend1 origin = (75,60) pct label = ("") mode = share;
legend2 origin = (75,55) pct label = ("") mode = share;
title1 "Figure 3.4 Rates and Points for 30 Year Fixed-Rate
Mortgage: Maximum-Average-Minimum";
```

```
title2 "1972 to 2012: with Rates and Points Displayed in
Different Y-axis";
title3 "Produced Using GPLOT";
FILENAME GSASFILE "&OUTLOC./Figure 3.4.&EXT";
  plot rate*year/haxis = axis1 vaxis = axis2 legend = legend1
    href = (1972 to 2012 by 5) LHREF = 2 des = "- Figure 3.4
Rates and Points in Different Y-axis";
  plot2 point*year/vaxis = axis3 legend = legend2;
  label rate = "Rate" point = "Point";
    run;
quit;

***************************************************************;
** Reproduce the same figures using sgplot                  *;
***************************************************************;
%LET OUTPUTFMT = PS;
ods listing gpath = "&outloc.";
ods graphics/reset = all width = 8in height = 6in noborder
OUTPUTFMT = &OUTPUTFMT. imagename = "FigSG 3_3";
ods proclabel = "Rates and Points for 30 Year Fixed-Rate
Mortgage";

title1 "Figure 3.3 Rates and Points for 30 Year Fixed-Rate
Mortgage: Maximum-Average-Minimum";
title2 "1972 to 2012: with Rates and Points Displayed in the
Same Y-axis";
title3 "Produced Using SGPLOT";
footnote1 "&pgmpth.";
proc sgplot data = ratepoint_sgplot des = "- Figure 3.3 Rates
and Points in the Same Y-axis";
  series x = year y = rate_avg/lineattrs = (thickness = 2
PATTERN = solid COLOR = black) MARKERS MARKERATTRS = (SYMBOL =
circlefilled COLOR = black) name = "rate" legendlabel =
"Rate";
  highlow x = year high = rate_max low = rate_min/lineattrs =
(thickness = 2 PATTERN = solid COLOR = black);
  series x = year y = point_avg/lineattrs = (thickness = 2
PATTERN = shortdash COLOR = red) MARKERS MARKERATTRS = (SYMBOL
= circle COLOR = red) name = "point" legendlabel = "Point";
  highlow x = year high = point_max low = point_min/lineattrs
= (thickness = 2 PATTERN = shortdash COLOR = red);

  xaxis VALUES = (1972 to 2012 by 5) label = "Year";
  yaxis VALUES = (0 to 20 by 1) label = "Rate and Point (%)";
  refline (1972 to 2012 by 5)/axis = x lineattrs = (pattern =
shortdash);
  keylegend "rate" "point"/noborder title = 'Legend:';
run;
quit;
```

```
ods graphics/reset = all width = 8in height = 6in noborder
OUTPUTFMT = &OUTPUTFMT. imagename = "FigSG 3_4";
ods proclabel = "Rates and Points for 30 Year Fixed-Rate
Mortgage";

title1 "Figure 3.4 Rates and Points for 30 Year Fixed-Rate
Mortgage: Maximum-Average-Minimum";
title2 "1972 to 2012: with Rates and Points Displayed in
Different Y-axis";
title3 "Produced Using SGPLOT";
footnote1 "&pgmpth.";
proc sgplot data = ratepoint_sgplot noautolegend des =
"- Figure 3.4 Rates and Points in Different Y-axes";
  series x = year y = rate_avg/lineattrs = (thickness = 2
PATTERN = solid COLOR = black) MARKERS MARKERATTRS = (SYMBOL =
circlefilled COLOR = black) curvelabel = "Rate";
  highlow x = year high = rate_max low = rate_min/lineattrs =
(thickness = 2 PATTERN = solid COLOR = black);
  series x = year y = point_avg/lineattrs = (thickness = 2
PATTERN = shortdash COLOR = red) y2axis MARKERS MARKERATTRS =
(SYMBOL = circle COLOR = red) curvelabel = "Point";
  highlow x = year high = point_max low = point_min/lineattrs
= (thickness = 2 PATTERN = shortdash COLOR = red) y2axis;

  xaxis VALUES = (1972 to 2012 by 5) label = "Year";
  yaxis VALUES = (0 to 20 by 1) label = "Rate (%)";
  y2axis VALUES = (0 to 3 by.5) label = "Point (%)";
  refline (1972 to 2012 by 5)/axis = x lineattrs = (pattern =
shortdash);
run;
quit;
```

4

Scatter and Jittered Scatter Plots

4.1 Introduction

A *scatter plot* is "a type of diagram using Cartesian coordinates to display values for two variables for a set of data. The data is displayed as a collection of points, each having the value of one variable determining the position on the horizontal axis and the value of the other variable determining the position on the vertical axis" (Wikipedia). Scatter plots are often used to visualize all individual data points, not just mean values as in line plots. In many circumstances, scatter plots might be preferred to line plots because they allow us to visualize all data points, including outliers. A scatter plot is a good example of letting the data speak for themselves, one important principle for data visualization and exploration (Tufte, 1983, 1997, 2006).

For scatter pots with discrete values in one axis, such as treatment group, gender, or age group, and continuous values in the other axis, the continuous values might be too close to each other to be separated or visually distinguished in the plots. A jittered scatter plot can be used in these scenarios. A *jittered scatter plot* is a type of scatter plot where the values on one axis (usually the discrete value at the horizontal) are randomly jittered away from each other within a range so the values in the other axis (usually continuous variables in the vertical axis) can be separated to be visually distinguished.

The sample scatter and jittered scatter plots are produced using both PROC GLOT and PROC SGPLOT procedures. The features, pros, and cons that are associated with the two procedures used to produce the scatter plots are discussed. SAS programs used to produce the sample figures are discussed and included in the Appendix (Section 4.6).

4.2 Application Examples

To illustrate the application and production of scatter and jittered scatter plots, three examples are presented in this chapter based on clinical research in the area of glaucoma therapy: the first example is a scatter plot of all subjects' IOP

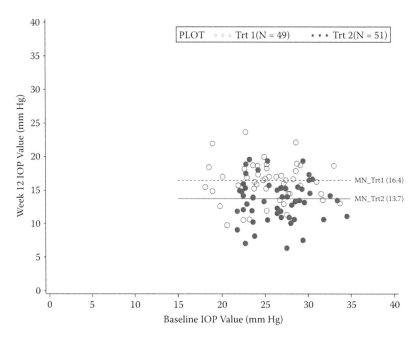

FIGURE 4.1
Week 12 compared to baseline IOP for two treatment groups.

values at the baseline and week 12 for two treatment groups with the mean value for each group displayed. The second example is similar to the first one, but with reference lines for various percent reductions added in the figure. The third example is a jittered scatter plot showing the change from baseline in all subjects' IOP values positioned side-by-side for the two treatment groups.

4.2.1 Example 1: Scatter Plots with Mean Value Displayed

A virtual clinical research on IOP reduction is simulated to randomize 100 subjects into two treatment groups ("TRT1" and "TRT2") at a 1:1 ratio. Subjects are simulated to have similar IOP values at the baseline but different at week 12 assuming different drug effects. A scatter plot displaying all 100 subjects' IOP values at the baseline and week 12 together with the mean IOP values at week 12 is shown in Figure 4.1. The figure demonstrates that the two treatment groups have similar IOP values at the baseline, but those in TRT2 have a lower IOP at week 12 than in TRT1, indicating better IOP reduction effect for TRT2.

4.2.2 Example 2: Scatter Plot with Mean Values and Reduction Reference Lines

It would also be interesting if we could count, directly in the figure, the number of subjects whose week 12 IOP is lower than the baseline, those with IOP

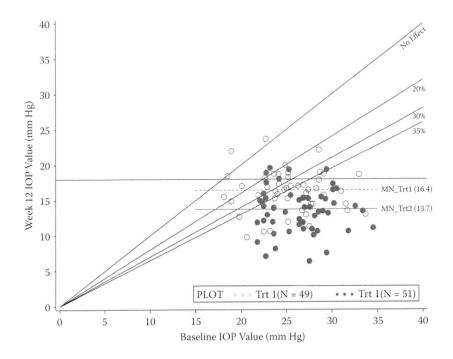

FIGURE 4.2
Week 12 compared to baseline IOP for two treatment groups with reference lines.

reduction achieving certain thresholds (e.g., 20%, 30%, and 35%), and those with Week 12 IOP of less than 18 mm Hg (a clinically meaningful value). Figure 4.2 is a more complicated scatter plot than Figure 14.1 displaying various reference lines (no effect, 20%, 30%, and 35% reduction and less than 18 mm Hg) together with the baseline and week 12 IOP values. The clinical folks like this type of plot because they can count, directly from the figure, how many subjects achieve a certain threshold IOP reduction (i.e., 20% reduction, etc.), and it is useful in responder analyses.

4.2.3 Example 3: Jittered Scatter Plots with IOP Values Displayed Side by Side

Both Figures 4.1 and 4.2 place the actual IOP values of the two treatment groups in the two axes (horizontal and vertical) using the same scale with treatment groups distinguished using different markers and colors. Figure 4.3 displays change from the baseline at week 12 and places the two treatment groups side by side in the horizontal axis. Treatment groups are assigned values of 1 and 2, and randomly jittered around 1 or 2 to allow separation for each individual subject's data. Because the jittering is only on the x-axis for the treatment group, it does not have any effect on the IOP values on the y-axis. Figure 4.3 is a commonly used jittered scatter plot.

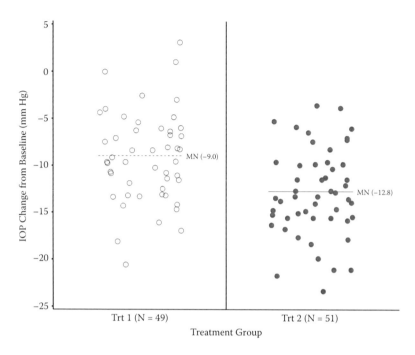

FIGURE 4.3

Change from baseline in IOP values for two treatment groups; randomly jittered scatter plot.

Figure 4.4 is a line-up jittered scatter plot that includes the same data points (IOP change) as in Figure 4.3, but presented in a more organized way for data visualization. Line-up jittered scatter plots are covered in detail in Chapter 5.

4.3 Producing the Sample Figures

4.3.1 Data Structure and SAS Annotated Dataset

Table 4.1 displays the data structure for one subject's IOP data. There are 100 subjects with a total of 100 records in the simulated dataset called IOP.

Three annotated datasets are created to display the reference lines and labels for mean values, percent reductions, and so on, in the sample figures using GPLOT. The first two annotated datasets are shown in Tables 4.2 and 4.3. The third annotated dataset is similar to the first one and is not shown.

4.3.2 Notes to SAS Programs

The SAS program that produced the three sample scatter and jittered scatter plots using the GPLOT and SGPLOT procedures is provided in the

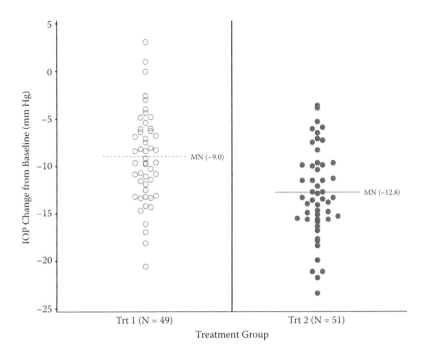

FIGURE 4.4
Change from baseline in IOP values for two treatment groups; line-up jittered scatter plot.

TABLE 4.1

Part of the Simulated IOP Data (Six Subjects' Data)

subjid	shuffle	trtgrp	iop_bsl	iop_w12
1001	0.739850279	2	23.6	10.5
1002	0.877109399	2	29.6	13.4
1003	0.277274282	1	18.5	18.7
1004	0.188775857	1	25.5	17.3
1005	0.752946201	2	25.3	19.6
1006	0.379947908	1	22.5	10.8

TABLE 4.2

Annotated Dataset Used to Draw Mean Value Reference Lines and Labels in Figure 4.1

FUNCTION	COLOR	XSYS	YSYS	HSYS	X	Y	SIZE	LINE	Text
MOVE	BLUE	2	2	4	15	16.4	1		
DRAW	BLUE	2	2	4	35	16.4	1	2	
label	BLUE	2	2	4	38	16.4	1		MN_Trt1(16.4)
MOVE	RED	2	2	4	15	13.74	1		
DRAW	RED	2	2	4	35	13.74	1	1	
label	RED	2	2	4	38	13.7	1		MN_Trt2(13.7)

TABLE 4.3

Annotated Dataset Used to Draw Various IOP Reduction and Mean Value
Reference Lines and Labels in Figure 4.2

FUNCTION	TEXT	SIZE	XSYS	YSYS	COLOR	X	Y	LINE	ANGLE
MOVE		1	2	2	BLACK	0	0		
DRAW		1	2	2	BLACK	40	40	1	
LABEL	No Effect	1	2	2	BLACK	38	38		45
MOVE		1	2	2	BLACK	0	0		
DRAW		1	2	2	BLACK	40	32	1	
LABEL	20%	1	2	2	BLACK	40	32		20
MOVE		1	2	2	BLACK	0	0		
DRAW		1	2	2	BLACK	40	28	1	
LABEL	30%	1	2	2	BLACK	40	28		30
MOVE		1	2	2	BLACK	0	0		
DRAW		1	2	2	BLACK	40	26	1	
LABEL	35%	1	2	2	BLACK	40	26		35
MOVE		1	2	2	BLUE	15	16.4		
DRAW		1	2	2	BLUE	35	16.4	2	
LABEL	MN_Trt1(16.4)	1	2	2	BLUE	38	16.4		0
MOVE		1	2	2	RED	15	13.7		
DRAW		1	2	2	RED	35	13.7	1	
LABEL	MN_Trt2(13.7)	1	2	2	RED	38	13.7		0

Appendix (Section 4.6). The following are the main sections and features of
the program.

4.3.2.1 Dataset Simulation

- A virtual clinical trial is simulated to include 100 subjects. Subjects
  are randomly assigned to either TRT 1 or TRT 2 at the 1:1 ratio. This
  is done by using an SAS function, *"ranuni(&seed.)"*, which generates
  random numbers between 0 and 1. If the random number is less
  than 0.5, the subject is assigned to TRT 1; otherwise, the subject is
  assigned to TRT 2.

- Subjects' IOP values are assigned based on a normal distribution
  with the preset mean and SD. Subjects' IOP values in the two treatment groups are set to be the same at the baseline, but different at
  week 12. Those in the TRT 2 group are assigned lower IOP values
  than those in the TRT 1 group, assuming TRT 2 is more effective
  than TRT 1 in IOP reduction.

4.3.2.2 Data Analyses and Manipulation

- The mean IOP values at week 12 of each treatment group are calculated using Proc Means with the results saved to a dataset *"iop_sum"*.

The subject number and mean IOP value for each treatment group are saved into macro variables using the function *call symputx()*.

- IOP data are manipulated to be separated for each treatment group and then merged together with reference values added to prepare a dataset (*IOP_ALL*) ready for plotting Figures 4.1 and 4.2.

- Mean IOP change from the baseline is calculated using *Proc Means* with the mean values saved in macro variables using the function *call symputx()*.

- The jittered dataset for the change from the baseline IOP: the treatment group values (1 for TRT 1 and 2 for TRT 2) are jittered by a random number from −0.25 to 0.25 so the subjects' data can be distinguished within each treatment group. The *STREAMINIT()* function is used to specify a seed value to use for subsequent random number generation by the *RAND* function. This way of generating random numbers in a specific range (e.g., −0.25 to 0.25) was suggested by Wicklin (2011).

4.3.2.3 SAS/GRAPH Annotate Datasets

- The SAS annotate macros *%ANNOMAC,%DCLANNO,* and *%LINE* are used to create three annotate datasets to produce reference lines for various IOP reduction and mean values in the sample figures using the GPLOT procedure. Section 3.3.2 in Chapter 3 has detailed descriptions of the use of the three annotate macros (SAS Institute Inc., 2012).

- ANNO_FIG1: used to draw reference lines for the mean values and display labels for the two treatment groups.

- ANNO_FIG2: used to draw various IOP reduction reference lines (no effect, 20%, 30%, and 35% reduction, and mean values) and labels in Figure 4.2.

- ANNO_FIG3: used to draw the mean value reference lines for the change from the baseline in Figure 4.3.

- In all the three annotated datasets, XSYS and YSYS are assigned to have value of "2" to use the data value for the (X,Y) coordinate system unit (SAS Institute Inc., 2012). The functions MOVE and DRAW are used to draw a straight line at the two (X,Y) coordinates. The function LABEL is used to label the TEXT in the specified (X,Y) coordinates. In Figure 4.2, the desired angles are selected to place the labels parallel to the reference lines. For example, "No Effect" is positioned beginning at the (38, 38) coordinates with the angle of 45 degrees. Note that the TEXT and ANGLE variables only work for the LABEL function and the LINE variable only works for the DRAW function.

4.3.2.4 Producing the Scatter Plots in GPLOT

- The *I = NONE* option is used in two SYMBOL statements to produce the scatters for the two treatment groups in different colors (blue and red) and symbols. Note that "albany amt/Unicode" is used in the FONT option with the value of "'25cb'x" representing open circles and "25cf'x" representing closed circles. These are the font-based markers and provide nicer and smoother (anti-aliased) markers than the symbol markers (*value = circle or dot*). A list of Unicodes for commonly used geometric shapes can be found at the link http://www.unicode.org/charts/PDF/U25A0.pdf.

```
SYMBOL1 H = 4 C = BLUE CO = BLUE I = NONE font = 'albany amt/
unicode' VALUE = '25cb'x;
SYMBOL2 H = 4 C = RED CO = RED I = NONE font = 'albany amt/
unicode' VALUE = '25cf'x;
```

- The OVERLAY option is used to place the two sets of the scatters for the two treatments on one set of axes in Figures 4.1 and 4.2. The RUN-group feature is used in one GPLOT procedure with two PLOT and two RUN statements.

```
PROC GPLOT DATA = IOP_ALL;
  PLOT iopw12_trt1*iopbsl_trt1 iopw12_trt2*iopbsl_trt2/
  overlay HAXIS = AXIS2 VAXIS = AXIS1 noframe legend
annotate = anno_fig1
  des = "Figure 4.1 Scatter Plot with Mean Values
Displayed";
  label  iopw12_trt1 = "Trt 1(N =%trim(&N_TRT1.))"
         iopw12_trt2 = "Trt 2(N =%trim(&N_TRT2.))";
RUN;

  PLOT iopw12_trt1*iopbsl_trt1 iopw12_trt2*iopbsl_trt2/
  overlay HAXIS = AXIS2 VAXIS = AXIS1 noframe legend
annotate = anno_fig2 vref = 18
  des = "Figure 4.2 Scatter Plot with Mean Values and
Reference Lines Displayed";
  label  iopw12_trt1 = "Trt 1(N =%trim(&N_TRT1.))"
         iopw12_trt2 = "Trt 2(N =%trim(&N_TRT2.))";
RUN;
```

- The *PLOT iop_chg * trtnum_jitter = trtgrp* statement is used to produce Figure 4.3 with the scatter plots for the two treatment groups displayed side by side. This is done in its own GPLOT procedure.

```
PROC GPLOT DATA = iopchg_jitter;;
  PLOT iop_chg * trtnum_jitter = trtgrp/HAXIS = AXIS2
VAXIS = AXIS1
  noframe nolegend href = 1.5 annotate = anno_fig3;
  format trtnum_jitter trtdf.;
  RUN;
```

4.3.2.5 Producing the Scatter Plots in SGPLOT

The same figures are reproduced using the SGPLOT procedure. The following are the main points.

- Datasets are manipulated to add variables X1, Y1 and X2, Y2 for drawing the mean values for each treatment group.
- The scatter plots are produced using the SCATTER statement. The appearance of the scatter plots and the labels are specified using the MARKERATTRS and LEGENDLABEL options.
- The reference lines for mean values are drawn using the Series plot statement with preset coordinates in the dataset (X1, Y1 and X2, Y2). The labels are displayed using the CURVELABEL option.
- The reference lines for no effect, 20%, 30%, and 35% reduction in Figure 4.2 are drawn using the LINEPARM plot statement in SGPLOT, which are also known as the parametric lines (Matange and Heath, 2011). A LINEPARM plot statement in SGPLOT is designed to draw a parametric line with an initial coordinate point and a slope. The labels are displayed using the CURVELABEL option.

```
proc sgplot data = iop_all noautolegend;
  scatter x = iopbsl_trt1 y = iopw12_trt1/markerattrs =
(symbol = circle color = blue size = 12) name = "trt1"
legendlabel = "TRT 1(N =%trim(&N_TRT1.))";
  scatter x = iopbsl_trt2 y = iopw12_trt2/markerattrs =
(symbol = circlefilled color = red size = 12) name =
"trt2" legendlabel = "TRT 2(N =%trim(&N_TRT2.))";

  ** identity and IOP reduction reference lines;
  lineparm x = 0 y = 0 slope = 1.0 /lineattrs = (color =
black pattern = solid) curvelabel = "No Effect"
curvelabelpos = max;
  lineparm x = 0 y = 0 slope = 0.8 /lineattrs = (color =
black pattern = solid) curvelabel = "20%" curvelabelpos =
max;
```

```
   lineparm x = 0 y = 0 slope = 0.7 /lineattrs = (color =
black pattern = solid) curvelabel = "30%" curvelabelpos =
max;
   lineparm x = 0 y = 0 slope = 0.65/lineattrs = (color =
black pattern = solid) curvelabel = "35%" curvelabelpos =
max;

   ** Lines for mean values
   series x = x1 y = y1/lineattrs = (color = blue pattern =
dash) curvelabel = "MN_Trt1(&mn_trt1.)" curvelabelpos = end;
   series x = x2 y = y2/lineattrs = (color = red pattern =
solid) curvelabel = "MN_Trt2(&mn_trt2.)" curvelabelpos =
end;
   xaxis values = (0 to 40 by 5) label = "Baseline IOP (mm
Hg)";
   yaxis values = (0 to 40 by 5) label = "Week12 IOP (mm
Hg)";
   refline 18/axis = y legendlabel = "18 mm Hg" labelloc =
inside labelpos = max lineattrs = (color = black pattern =
solid);
   keylegend "trt1" "trt2"/noborder;
run;
```

4.3.2.6 *Producing Line-Up Jittered Scatter Plot*

A line-up scatter plot is introduced and shown in Figure 4.4, which has the same data points as Figure 4.3, but displayed in a more organized and better visualized way. In Figure 4.4, the data points (IOP change) are lined-up based on their values rather than randomly jittered as in Figure 4.3. The line-up of the data points is done in the macro %LineUp_Jitter(). Detailed introduction and production of line-up jittered scatter plots are covered in Chapter 5.

4.4 Summary and Discussion

Scatter plots are produced using both GPLOT and SGPLOT procedures in the chapter. When producing the reference line for different reductions (no effect or 0%, 20%, 30%, and 35%), the default reference statements (VREF or HREF in GPLOT and REFLINE in SGPLOT) cannot be used. This reference line is produced by using the annotate facility in the GPLOT procedure and the LINEPARM plot statement in the SGPLOT procedure.

Depending on the nature or requirement of the scatter plots, one procedure might be easier to use than the other. However, the statements and settings

TABLE 4.4

Comparing PROC GPLOT and PROC SGPLOT in Producing Scatter Plots

Features	GPLOT	SGPLOT
Scatter plot	Using *INTERPOL = none* in SYMBOL statement	Using SCATTER plot statement *Scatter x = y =;*
Scatter and marker attributes	Using SYMBOL statement	Using LINEATTRS and MARKERATTRS plot statements
Axis attributes	AXIS global statements and VAXIS/HAXIS plot statement options	XAXIS and YAXIS plot statements
Legend	Global LEGEND statement	LEGENDLABEL in MARKERS plot statement
Reference lines and labels for mean values	Using the annotate facility	Using SERIES plot statement
Reference lines and labels for percent reduction	Using the annotate facility	Using LINEPARM plot statement with an initial point and a slope
Pros	Very high flexibility in using the annotate facility to draw the various reference lines and place labels, including placing the labels at different angles.	Easy to draw the percent reduction reference lines and place labels using LINEPARM plot statement with an original point and a slope and the CURVELABEL option.
Cons	Have to learn annotate facility.	Not as flexible; cannot easily place the reference labels with angles.

used to produce scatter plots in GPLOT and SGPLOT are quite different. Table 4.4 summarizes the main features in producing the scatter plots using the two procedures.

4.5 References

Matange, S., and Heath, D. 2011. *Statistical Graphics Procedures by Example: Effective Graphs Using SAS®*. Cary, NC: SAS Institute Inc.

SAS Institute Inc. 2012. *SAS/GRAPH® 9.3: Reference*, 3rd ed. Cary, NC: SAS Institute Inc.

Tufte, E. 1983. *The Visual Display of Quantitative Information*. Cheshire, CT: Graphics Press.

Tufte, E. 1997. *Visual Explanations*. Cheshire, CT: Graphics Press.

Tufte, E. 2006. *Beautiful Evidence*. Cheshire, CT: Graphics Press.

Wicklin, R. 2011. "How to Generate Random Numbers in SAS," SAS Blogs, http://blogs. sas.com/content/iml/2011/08/24/how-to-generate-random-numbers-in-sas/.

Wikipedia. "Scatter Plot," http://en.wikipedia.org/wiki/Scatter_plot.

4.6 Appendix: SAS Programs for Producing the Sample Figures

```
***************************************************************;
* Program Name: Scatter and Jittered Scatter Plots.sas       *;
* Descriptions: Producing the following sample figures in     *;
* chapter 4 Using GPLOT and SGPLOT                            *;
* - Figure 4.1 Baseline and Week 12 IOP Values for Two        *;
* Treatment Groups:                                           *;
*       -. Scatter Plot with Mean Values                      *;
* - Figure 4.2 Baseline and Week 12 IOP Values for Two        *;
* Treatment Groups:                                           *;
*       -. Scatter Plot with Mean Values and Reference Lines  *;
* - Figure 4.3 Change from Baseline in IOP Values for Two     *;
* Treatment Groups:                                           *;
*       -. Jittered Scatter Plot with Mean Values             *;
* - Figure 4.4 Line-up Jittered Scatter Plot                  *;
***************************************************************;
options mprint symbolgen nodate nonumber validvarname = v7
orientation = landscape;
%let pgmname = Chapter 4 Scatter and Jittered Scatter Plots.
sas;
%let pgmloc = C:\SASBook\SAS Programs;
%let outloc = C:\SASBook\Sample Figures\Chapter 4;
%let pgmpth = &pgmloc.\&pgmname. &sysdate9. &systime. SAS
V&sysver.;


** Set-up the site, subject number and SD for data simulation;
%let subjnum = 100;
%LET SD = 3.5;
%LET SEED = 04;


** Generate the required number of subjects;
data IOP;
  do i = 1 to &subjnum.;
    subjid = 1000+ i;
    shuffle = ranuni (&SEED.);
      if 0 < = shuffle < 0.5 then do;
        trtgrp = 1;
        iop_bsl = round((RANNOR(&SEED.)* &SD. + 25),.1);
** BSL IOP for TRT 1;
        iop_w12 = round((RANNOR(&SEED.)* &SD. + 16),.1);
** Wk12 IOP for TRT 1;
      end;
      if 0.5 < = shuffle < = 1 then do;
        trtgrp = 2;
        iop_bsl = round((RANNOR(&SEED.)* &SD. + 25),.1);
** BSL IOP for TRT 2;
```

```
            iop_w12 = round((RANNOR(&SEED.) * &SD. + 14),.1);
** WK12 IOP for TRT 2;
        end;
        output;
   end;
   drop i;
run;

** Get the mean IOP values of the 2 treatments at week 12;
proc means data = iop noprint;
   class trtgrp;
   var iop_w12;
   output out = iop_sum n = n mean = mn;
run;

** Save the mean and N of each treatment group into macro
variables;
data _null_;
   set iop_sum;
   call symputx('MN_Trt'||left(put(trtgrp,best.)), put(mn, 4.1));
   call symputx('N_Trt'||left(put(trtgrp,best.)), put(n, 3.0));
run;

proc format;
   value trtdf
      1 = "Trt 1(N =%trim(&N_TRT1.))"
      2 = "Trt 2(N =%trim(&N_TRT2.))"
      OTHER = " ";
run;

data iop_trt1;
   set iop;
   where trtgrp = 1;
   rename subjid = subjid_trt1 iop_bsl = iopbsl_trt1 iop_w12 =
iopw12_trt1;
   drop shuffle trtgrp;
run;

data iop_trt2;
   set iop;
   where trtgrp = 2;
   rename subjid = subjid_trt2 iop_bsl = iopbsl_trt2 iop_w12 =
iopw12_trt2;
   drop shuffle trtgrp;
run;

** IOP for 2 TRTs;
data iop_trt;
   merge iop_trt1 iop_trt2;
run;
```

```
data iop_all;
  set iop_trt;
  x1 = 15; y1 = &mn_trt1.;
  x2 = 15; y2 = &mn_trt2.;
  output;
  x1 = 35; y1 = &mn_trt1.; ** mean line for trt1;
  x2 = 35; y2 = &mn_trt2.; ** mean line for trt2;
  output;
run;

** Change from baseline;
data iop_chg;
  set iop;
  iop_chg = iop_w12 - iop_bsl;
run;

* Summary Statistics for IOP change from baseline by treatment
group;
PROC MEANS DATA = iop_chg NWAY MEAN STD STDERR MIN MAX MEDIAN
maxdec = 2 noprint;
  CLASS trtgrp;
  VAR iop_chg;
  OUTPUT OUT = iopchg_sum MEAN = Mean;
RUN;

data _null_;
  set iopchg_sum;
    call symputx('CHGMN_Trt'||left(put(trtgrp,best.)),
put(mean, 5.1));
run;

** Jitter the treatment group in x-axis by 0.25 unit;
* 0.75 to 1.25 for TRT 1 and 1.75 to 2.25 for TRT 2;
data iopchg_jitter;
  set iop_chg;
  call streaminit(123);
  a = -0.25; b = 0.25;
  u = rand("Uniform"); ** U[0,1];
  jitter = a + (b-a)*u; ** U[a,b];
  trtnum_jitter = trtgrp + jitter;
run;

* Make the SAS Annotate dataset macros available for use;
%ANNOMAC;
* Reference lines for Week 12 IOP mean values in sample
figure 1;
DATA anno_fig1;
  SET iop_sum;
  %DCLANNO;
  RESULT_MN = round(mn,.01);
```

```
  RETAIN SEMULT 1 NUM_OFFSET.1 SHIFT_VAL.25;
  SIZE = 1; XSYS = '2'; YSYS = '2';

  %* draw the horizontal line for mean *;
  if trtgrp = 1 then do;
    %LINE(15, RESULT_MN, 35, RESULT_MN, BLUE, 2, SIZE);
    x = 38; y = &mn_trt1.; function = 'label'; text =
"MN_Trt1(&mn_trt1.)";
    output;
  end;
  if trtgrp = 2 then do;
    %LINE(15, RESULT_MN, 35, RESULT_MN, RED, 1, SIZE);
    x = 38; y = &mn_trt2.; function = 'label'; text =
"MN_Trt2(&mn_trt2.)";
    output;
  end;
  KEEP X Y FUNCTION COLOR SIZE LINE TEXT HSYS XSYS YSYS;
RUN;

** Reference lines and labels to be used for annotation in
sample figure 2;
data anno_fig2;
  length function $8 text $15.;
  SIZE = 1; XSYS = '2'; YSYS = '2';

  %LINE(0, 0, 40, 40, BLACK, 1, SIZE); ** No Effect Ref Line;
  x = 38; y = 38; angle = 45; function = 'LABEL';
  text = "No Effect"; output;

  %LINE(0, 0, 40, 32, BLACK, 1, SIZE); ** 20% Reduction Ref Line;
  x = 40; y = 32; angle = 20; function = 'LABEL';
  text = "20%"; output;

  %LINE(0, 0, 40, 28, BLACK, 1, SIZE); ** 30% Reduction Ref Line;
  x = 40; y = 28; angle = 30; function = 'LABEL';
  text = "30%"; output;

  %LINE(0, 0, 40, 26, BLACK, 1, SIZE); ** 35% Reduction Ref Line;
  x = 40; y = 26; angle = 35; function = 'LABEL';
  text = "35%"; output;

  %LINE(15, &mn_trt1., 35, &mn_trt1., BLUE, 2, SIZE);
** Ref Line for Trt1 Mean;
  x = 38; y = &mn_trt1.; angle = 0; function = 'LABEL';
  text = "MN_Trt1(&mn_trt1.)"; output;

  %LINE(15, &mn_trt2., 35, &mn_trt2., RED, 1, SIZE);
** Ref Line for Trt2 Mean;
  x = 38; y = &mn_trt2.; angle = 0; function = 'LABEL';
  text = "MN_Trt2(&mn_trt2.)"; output;
run;
```

```
data anno_fig2;
   set anno_fig2;
   if function ne "LABEL" then DO; ** TEXT and ANGLE only apply
to LABEL function;
      text = " "; ANGLE =.;
   END;
   if function ne "DRAW" then line =.; ** LINE variable only
apply to DRAW function;
run;

* Reference lines for change from baseline at Week 12 mean
values in sample figure 3;
DATA anno_fig3;
   SET iopchg_sum;
   RESULT_MN = round(mean,.01);
   RETAIN SEMULT 1 NUM_OFFSET.1 SHIFT_VAL.25;
   SIZE = 1; XSYS = '2'; YSYS = '2';
   * Draw the horizontal line for mean *;
   if trtgrp = 1 then do;
      %LINE(trtgrp-SHIFT_VAL, RESULT_MN, trtgrp+SHIFT_VAL,
RESULT_MN, BLUE, 2, SIZE);
      x = 1.38; y = RESULT_MN; function = 'LABEL';
      text = "MN (" || put(RESULT_MN, 5.1) || ")";
      output;
   end;
   if trtgrp = 2 then do;
      %LINE(trtgrp-SHIFT_VAL, RESULT_MN, trtgrp+SHIFT_VAL,
RESULT_MN, RED, 1, SIZE);
      x = 2.38; y = RESULT_MN; function = 'LABEL';
      text = "MN (" || put(RESULT_MN, 5.1) || ")";
      output;
   end;
   KEEP X Y FUNCTION COLOR FUNCTION TEXT SIZE LINE HSYS XSYS YSYS;
RUN;

%LET FONTNAME = Times;
%LET DRIVER = PSCOLOR;%LEt EXT = PS;
goptions
   reset    = all
   GUNIT    = PCT
   rotate   = landscape
   gsfmode  = replace
   gsfname  = GSASFILE
   device   = &DRIVER
   lfactor  = 1
   hsize    = 8 in
   horigin  = 0 in
   vsize    = 6.5 in
   vorigin  = 0 in
   ftext    = "&FONTNAME"
```

```
  htext    = 10pt
  ftitle   = "&FONTNAME"
  htitle   = 10pt
;

SYMBOL1 H = 4 C = BLUE     CO = BLUE I = NONE font = 'albany
amt/unicode' VALUE = '25cb'x;
SYMBOL2 H = 4 C = RED      CO = RED I = NONE font = 'albany
amt/unicode' VALUE = '25cf'x;
AXIS1 OFFSET = (1,1) ORDER = (0 to 40 by 5) LABEL = (FONT =
"&FONTNAME" h = 2.5 ANGLE = 90 "Week 12 IOP Value (mm Hg)")
VALUE = (H = 2.5) MINOR = NONE;
AXIS2 OFFSET = (1,1) ORDER = (0 to 40 by 5) LABEL = (FONT =
"&FONTNAME" h = 2.5 "Baseline IOP Value (mm Hg)") VALUE =
(H = 2.5) MINOR = NONE;

FILENAME GSASFILE "&OUTLOC.\Figure 4.1.&EXT.";
title1 "Figure 4.1 Baseline and Week 12 IOP Values for Two
Treatment Groups";
title2 "Scatter Plot with Mean Values";
footnote1 "&pgmpth.";
ods proclabel = "Baseline and Week 12 IOP Values for Two
Treatment Groups";
PROC GPLOT DATA = IOP_ALL;
  PLOT iopw12_trt1*iopbsl_trt1 iopw12_trt2*iopbsl_trt2/
    overlay HAXIS = AXIS2 VAXIS = AXIS1 noframe legend
annotate = anno_fig1
    des = "Figure 4.1 Scatter Plot with Mean Values
Displayed";
  label iopw12_trt1 = "Trt 1(N =%trim(&N_TRT1.))" iopw12_trt2
= "Trt 2(N =%trim(&N_TRT2.))";
RUN;

FILENAME GSASFILE "&OUTLOC.\Figure 4.2.&EXT.";
title1 "Figure 4.2 Baseline and Week 12 IOP Values for Two
Treatment Groups";
title2 "Scatter Plot with Mean Values and Reference Lines";
footnote1 "&pgmpth.";
  PLOT iopw12_trt1*iopbsl_trt1 iopw12_trt2*iopbsl_trt2/
    overlay HAXIS = AXIS2 VAXIS = AXIS1 noframe legend
annotate = anno_fig2 vref = 18
    des = "Figure 4.2 Scatter Plot with Mean Values and
Reference Lines Displayed";
  label  iopw12_trt1 = "Trt 1(N =%trim(&N_TRT1.))"
         iopw12_trt2 = "Trt 2(N =%trim(&N_TRT2.))";
RUN;

AXIS1 OFFSET = (1,1) ORDER = (-25 to 5 by 5) LABEL = (FONT =
"&FONTNAME" h = 2.5 ANGLE = 90 "IOP Change from Baseline
(mm Hg)") VALUE = (H = 2.5) MINOR = NONE;
```

```
AXIS2 major = none OFFSET = (1,1) ORDER = (0.5 to 2.5 by 0.5)
LABEL = (FONT = "&FONTNAME" h = 2.5 "Treatment Group") VALUE =
(H = 2.5) MINOR = NONE;

FILENAME GSASFILE "&OUTLOC.\Figure 4.3.&EXT.";
title1 "Figure 4.3 Change from Baseline in IOP Values for Two
Treatment Groups";
title2 "Jittered Scatter Plot with Mean Values";
footnote1 "&pgmpth.";
ods proclabel = "Change from Baseline in IOP Values";
PROC GPLOT DATA = iopchg_jitter;;
  PLOT iop_chg * trtnum_jitter = trtgrp/HAXIS = AXIS2 VAXIS =
AXIS1 noframe nolegend href = 1.5 annotate = anno_fig3;
  format trtnum_jitter trtdf.;
RUN;
QUIT;

** Prepare the dataset to be used for SGPLOT;
data iopchg_trt1;
  set iopchg_jitter;
  where trtgrp = 1;
  rename subjid = subjid_trt1 iop_chg = iopchg_trt1
trtnum_jitter = trt1_jitter;
  keep subjid iop_chg trtnum_jitter;
run;

data iopchg_trt2;
  set iopchg_jitter;
  where trtgrp = 2;
  rename subjid = subjid_trt2 iop_chg = iopchg_trt2
trtnum_jitter = trt2_jitter;
  keep subjid iop_chg trtnum_jitter;
run;

data iopchg_both;
  merge iopchg_trt1 iopchg_trt2;
run;

** Data to draw mean lines;
data iopchg_line;
  x1 = 0.75; y1 = &CHGMN_TRT1.;
  x2 = 1.75; y2 = &CHGMN_TRT2.;
  output;
  x1 = 1.25; y1 = &CHGMN_TRT1.; ** mean line for change in trt1;
  x2 = 2.25; y2 = &CHGMN_TRT2.; ** mean line for change in trt2;
  output;
run;
```

```
data iopchg_all;
  merge iopchg_both iopchg_line;
run;

%LET OUTPUTFMT = PS;
ods listing gpath = "&outloc.";
ods graphics/reset = all width = 8in height = 6.5in noborder
OUTPUTFMT = &OUTPUTFMT. imagename = "FigSG 4_1";
title1 "Figure 4.1 Baseline and Week 12 IOP Values for Two
Treatment Groups";
title2 "Scatter Plot with Mean Values";
footnote1 "&pgmpth.";
proc sgplot data = iop_all noautolegend;
  scatter x = iopbsl_trt1 y = iopw12_trt1/markerattrs =
(symbol = circle color = blue size = 12) name = "trt1"
legendlabel = "TRT 1(N =%trim(&N_TRT1.))";
  scatter x = iopbsl_trt2 y = iopw12_trt2/markerattrs =
(symbol = circlefilled color = red size = 12) name = "trt2"
legendlabel = "TRT 2(N =%trim(&N_TRT2.))";
  ** Lines for mean values;
  series x = x1 y = y1/lineattrs = (color = blue pattern =
dash) curvelabel = "MN_Trt1(&mn_trt1.)" curvelabelpos = end;
  series x = x2 y = y2/lineattrs = (color = red pattern =
solid) curvelabel = "MN_Trt2(&mn_trt2.)" curvelabelpos = end;
  xaxis values = (0 to 40 by 5) label = "Baseline IOP (mm Hg)";
  yaxis values = (0 to 40 by 5) label = "Week12 IOP (mm Hg)";
  keylegend "trt1" "trt2"/noborder;
run;
quit;

ods graphics/reset = all width = 8in height = 6.5in noborder
OUTPUTFMT = &OUTPUTFMT. imagename = "FigSG 4_2";
title1 "Figure 4.2 Baseline and Week 12 IOP Values for Two
Treatment Groups";
title2 "Scatter Plot with Mean Values and Reference Lines";
footnote1 "&pgmpth.";
proc sgplot data = iop_all noautolegend;
  scatter x = iopbsl_trt1 y = iopw12_trt1/markerattrs =
(symbol = circle color = blue size = 12) name = "trt1"
legendlabel = "TRT 1(N =%trim(&N_TRT1.))";
  scatter x = iopbsl_trt2 y = iopw12_trt2/markerattrs =
(symbol = circlefilled color = red size = 12) name = "trt2"
legendlabel = "TRT 2(N =%trim(&N_TRT2.))";

  ** Identity and IOP reduction reference lines;
  lineparm x = 0 y = 0 slope = 1.0 /lineattrs = (color = black
pattern = solid) curvelabel = "No Effect" curvelabelpos = max;
```

```
   lineparm x = 0 y = 0 slope = 0.8 /lineattrs = (color = black
pattern = solid) curvelabel = "20%" curvelabelpos = max;
   lineparm x = 0 y = 0 slope = 0.7 /lineattrs = (color = black
pattern = solid) curvelabel = "30%" curvelabelpos = max;
   lineparm x = 0 y = 0 slope = 0.65/lineattrs = (color = black
pattern = solid) curvelabel = "35%" curvelabelpos = max;

   ** Lines for mean values;
   series x = x1 y = y1/lineattrs = (color = blue pattern =
dash) curvelabel = "MN_Trt1(&mn_trt1.)" curvelabelpos = end;
   series x = x2 y = y2/lineattrs = (color = red pattern =
solid) curvelabel = "MN_Trt2(&mn_trt2.)" curvelabelpos = end;
   xaxis values = (0 to 40 by 5) label = "Baseline IOP (mm Hg)";
   yaxis values = (0 to 40 by 5) label = "Week12 IOP (mm Hg)";
   refline 18/axis = y legendlabel = "18 mm Hg" labelloc = inside
labelpos = max lineattrs = (color = black pattern = solid);
   keylegend "trt1" "trt2"/noborder;
run;
quit;

** ods listing gpath = "&outloc.";
ods graphics/reset = all width = 8in height = 6.5in noborder
OUTPUTFMT = &OUTPUTFMT. imagename = "FigSG 4_3";
title1 "Figure 4.3 Change from Baseline in IOP Values for Two
Treatment Groups";
title2 "Jittered Scatter Plot with Mean Values";
footnote1 "&pgmpth.";
proc sgplot data = iopchg_all noautolegend;
   scatter x = trt1_jitter y = iopchg_trt1/markerattrs =
(symbol = circle color = blue size = 12);
   scatter x = trt2_jitter y = iopchg_trt2/markerattrs =
(symbol = circlefilled color = red size = 12);
   ** Lines for mean values;
   series x = x1 y = y1/lineattrs = (color = blue pattern =
dash) curvelabel = " MN(&CHGMN_TRT1.)" curvelabelpos = end;
   series x = x2 y = y2/lineattrs = (color = red pattern =
solid) curvelabel = " MN(&CHGMN_TRT2.)" curvelabelpos = end;
   xaxis values = (0.5 to 2.5 by.5) label = "Treatment Group"
display = (NOTICKS);
   yaxis values = (-30 to 10 by 5) label = "IOP Change from
Baseline (mm Hg)" display = (NOTICKS);

refline 1.5/axis = x lineattrs = (color = black pattern = dash);
   format trt1_jitter trt2_jitter trtdf.;
   keylegend "trt1" "trt2"/noborder;
run;
quit;
```

```
** Produce the Line-up jittered scatter plot;
%macro LineUp_Jitter (idn =, xvar =, yvar =, lvl =, jitter =,
odn =);
** Get the Min and Max values and save them into macro
variables;
PROC MEANS DATA = &idn. MIN MAX maxdec = 2 noprint;
  VAR &yvar.;
  OUTPUT OUT = tmp_STAT min = min max = max;
RUN;
data _null_;
  set tmp_stat;
  call symput ('Min', min);
  call symput ('Max', max);
run;

proc sort data = &idn. out = d_tmp;
  by &yvar.;
run;

data add_tmp;
  set d_tmp;
  do i = floor(&min.) to ceil(&max.) by &lvl.;
    if i < = &yvar. < i+1 then yvar_int = i;
  end;
run;

** get the number of data points at each interval by x-axis
value;
proc freq data = add_tmp noprint;
  tables &xvar.*yvar_int/out = freqdata;
run;

data jitter_sum;
  retain pos_neg 1;
  set freqdata;
  pos_neg = 1;

  if count = 1 then do;
    xvar_j = &xvar.;
    output;
  end;

  if count>1 then do;
    if mod(count,2) > 0 then do; ** odd count number;
      xvar_j = &xvar.; pos = &xvar.; neg = &xvar.; output;
      do i = 1 to count-1 by 1;
        if pos_neg = 1 then do;
          xvar_j = neg-&jitter;
            neg = xvar_j;
        end;
```

```
         else do;
            xvar_j = pos+&jitter;
               pos = xvar_j;
         end;
         output;
         pos_neg = -1*pos_neg;
      end;
   end;

   if mod(count,2) = 0 then do; ** even count number;
      pos = &xvar.; neg = &xvar.;
      do i = 1 to count by 1;
         if pos_neg = 1 then do;
            xvar_j = neg-&jitter/2;
               neg = xvar_j-&jitter/2;
         end;
         else do;
            xvar_j = pos+&jitter/2;
               pos = xvar_j+&jitter/2;
         end;
         output;
         pos_neg = -1*pos_neg;
      end;
   end;
  end;
run;

proc sort data = add_tmp;
  by yvar_int &xvar.;
run;

proc sort data = jitter_sum;
  by yvar_int &xvar.;
run;

data jitter_sum;
  set jitter_sum;
  by yvar_int &xvar.;
  ord = _n_;
run;

data add_tmp;
  set add_tmp;
  by yvar_int &xvar.;
  ord = _n_;
run;

** Output dataset: Jittered dataset for ploting;
data &odn.;
    merge jitter_sum (in = a) add_tmp;
```

```
      by ord;
      if a;
      keep &yvar. &xvar. yvar_int xvar_j;
%mend LineUp_Jitter;

%LineUp_Jitter (idn = iop_chg, yvar = iop_chg, xvar = trtgrp,
lvl = 0.5, jitter = 0.06, odn = IOPCHG_LJ);

%LET FONTNAME = Times;
%LET DRIVER = PSCOLOR;%LET EXT = PS;
%LET FMTTITLE = H = 2.5 JUSTIFY = CENTER FONT = "&FONTNAME";
%LET FMTFOOT = H = 2 JUSTIFY = CENTER FONT = "&FONTNAME";

SYMBOL1 H = 4 C = BLUE     CO = BLUE I = NONE font = 'albany
amt/unicode' VALUE = '25cb'x;
SYMBOL2 H = 4 C = RED      CO = RED I = NONE font = 'albany
amt/unicode' VALUE = '25cf'x;
AXIS1 OFFSET = (1,1) ORDER = (-25 to 5 by 5) LABEL = (FONT =
"&FONTNAME" h = 2.5 ANGLE = 90 "IOP Change from Baseline
(mm Hg)") VALUE = (H = 2.5) MINOR = NONE;
AXIS2 major = none OFFSET = (1,1) ORDER = (0.5 to 2.5 by 0.5)
LABEL = (FONT = "&FONTNAME" h = 2.5 "Treatment Group") VALUE =
(H = 2.5) MINOR = NONE;

FILENAME GSASFILE "&OUTLOC.\Figure 4.4.emf.";
title1 "Figure 4.4 Change from Baseline in IOP Values for Two
Treatment Groups";
title2 "Line-up Jittered Scatter Plot with Mean Values";
footnote1 "&pgmpth.";
PROC GPLOT DATA = iopchg_lj;
  PLOT iop_chg * xvar_j = trtgrp
    /HAXIS = AXIS2 VAXIS = AXIS1 noframe nolegend href = 1.5
annotate = anno_fig3;
  label trtgrp = 'Legend:'; format xvar_j trtdf.;
RUN;
QUIT;
```

5

Line-Up Jittered Scatter Plots

5.1 Introduction

A line-up jittered scatter plot, also known as a bee swarm plot, is displayed in Figure 4.4 of Chapter 4, which provides a more organized and better data visualization than its randomly jittered counterpart in Figure 4.3. A line-up jittered scatter plot or bee swarm plot is a new type of jittered scatter plot where the data points with differences within a certain limit or range on one axis (usually the y-axis) are lined-up together to be positioned in the same group. The data points within the same line-up group sometimes can also be reassigned to have the same value for better data visualization. The certain limit or the range of the data depends on the nature of the data and normally the difference within the same line-up group is small enough to be neglected or not meaningful; for example, 0.5 mm Hg for intraocular pressure (IOP) values in glaucoma clinical trials or 0.15 kg in body weight measurement, and so on.

Line-up jittered scatter plots provide a more organized way for data to be visualized and displayed. It is also useful when we need to find out how many data points are within a certain range on the y-axis.

5.2 Algorithms for Line-Up Jittered Scatter Plots

The following are the detailed algorithms needed to jitter data points by the line-up groups to prepare a dataset that is ready for a line-up jittered scatter plot.

- Determine the maximum and the minimum values of the data used for the plot.
- Choose a "similar" value for a line-up group for one axis (usually the y-axis) and a jittering level for another axis (usually the x-axis): the value chosen for the line-up group should be small enough not to affect the true meaning of the data. The jittered level might require experimentation—try and see—before the final value is selected.

- Beginning at the minimum endpoint, align the data points whose difference from each other is within the "similar" value for line-up to the same group.
- Count the number of data points within each line-up group on one axis (usually the y-axis) and decide the jittering technique for each group based on the number of data counts at the group on another axis (usually the x-axis):
  - If the data count is only one, no jitter is needed.
  - If the data count is an odd number, do not jitter the first number, then jitter one data point in the left side and another in the right side by the jittering level, and continue this way until all data points in the group are jittered.
  - If the data count is an even number, jitter the first number half of the jittering level to the left, the second number half the jittering level to the right, then jitter one data point to the left side and another to the right side by the jittering level, and continue this way until all data points in the group are jittered.
- Reassign the same value to all of the data points within the same line-up group. This step is optional and extreme caution should be taken because this step will change the original data values. Do this only for a special need or when the differences within the same line-up group are so small that they can be ignored or neglected.
- Plot the data points using the original value or the reassigned value (for those within the same group) by the line-up jittered value.

The macro *LineUp_Jitter()* included in the SAS programs in Chapters 4 and 5 is written using the above algorithm.

5.3 Application Examples

To illustrate the application and demonstrate the production of line-up jittered scatter plots, two sample plots are presented in this chapter. The sample figures, based on clinical research in the glaucoma therapeutic area, display individual IOP values on the y-axis by treatment group and eye on the x-axis. We are interested to see the actual IOP values for each eye (OD and OS) with the mean values displayed for all subjects by treatment group. Further, we would like to see the IOP values placed in a figure ranging from the smallest to the largest by the 0.5 mm Hg interval. A line-up jittered scatter plot is a good visualization tool to meet these needs. In order to place the data points on the x-axis at the appropriate level and be able to distinguish among them but not have them be too close or too far away from each other,

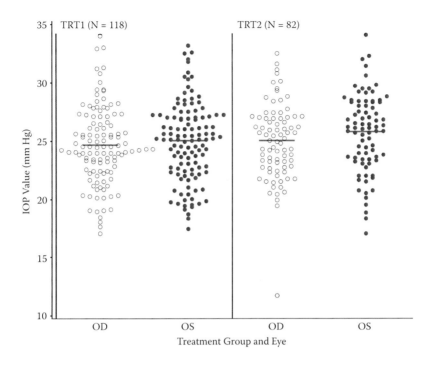

FIGURE 5.1
Line-up jittered scatter plot for IOP by treatment and eye; IOP data on the y-axis are the original values.

it is often necessary to experiment in order to select the right jitter level. In the sample figures, the IOP values are jittered far away to be distinguished among each other but still within their treatment and eye group boundaries.

Figure 5.1 uses the actual IOP values while Figure 5.2 uses the reassigned values on the y-axis. Both figures use the same jittering level on the x-axis. We need to be very careful when using the reassigned values on the y-axis to make sure the meaning of the data is not lost, and this is normally not recommended in clinical data analyses and visualization unless it is required and agreed to by all parties. Figure 5.1 is the preferred model for real data publication, but Figure 5.2 might be better for counting the data points ranging from the smallest to the largest at the 0.5 mm Hg interval.

5.4 Producing the Sample Figures

5.4.1 Data Structure and SAS Annotated Dataset

A dataset is simulated to include the IOP data for 200 subjects for both eyes (right eye [OD] and left eye [OS]) from a normal distribution with the mean

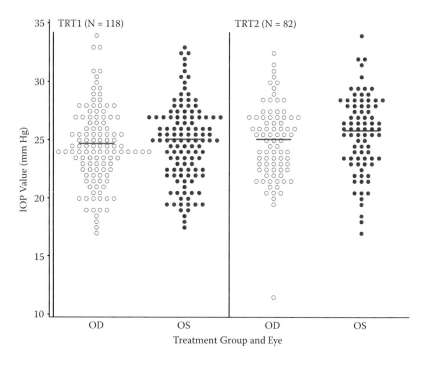

FIGURE 5.2
Line-up jittered scatter plot for IOP by treatment and eye; IOP values within the same line-up group on the y-axis might differ from each other by 0.5 mm Hg.

value of 25 mm Hg and the standard deviation (SD) of 3.5 mm Hg. Subjects are randomly assigned to group TRT1 or TRT2 at the 1:1 ratio. A variable *index* is assigned to have the values 1, 2, 3, and 4 to represent the 2 eyes in the 2 treatment groups. Part of the simulated dataset is shown in Table 5.1. There are 200 subjects with a total of 400 records in the simulated dataset.

The SAS/GRAPH annotate facility is used to display the mean IOP value for each eye within each treatment group in the sample figure in the GPLOT procedure. The annotated dataset is displayed in Table 5.2. Variable Y contains the mean IOP values for the 2 eyes in the 2 treatment groups, which is the location where the line representing the mean value need to be drawn on the y-axis. The midpoint for each treatment/eye is positioned at 1, 2, 3, and 4 on the x-axis, and the mean line starts 0.2 unit space left of the midpoint and extends to 0.2 unit space right of the midpoint. For example, the mean line for the OD of TRT1 group starts at 0.8 and ends at 1.2. By using the *MOVE* and *DRAW* functions in the annotate facility, SAS first moves the point to the coordinate (0.8, 24.69) and then draws a line to the coordinate point (1.2, 24.69). For a more detailed introduction of SAS annotate facility, please read Part 5 in the *SAS/GRAPH 9.3 Reference* (SAS Institute Inc., 2012).

In producing the sample figures using SGPLOT, SG annotation is used to display the mean value in a short line for each eye within each treatment

TABLE 5.1

Part of the Simulated Dataset to Produce the Line-Up Jittered Scatter Plot in Example 1

SUBJID	EYE	TRT	IOP	Index
1001	OD	TRT2	27	3
1001	OS	TRT2	20.8	4
1002	OD	TRT2	23.4	3
1002	OS	TRT2	22.1	4
1003	OD	TRT1	24.6	1
1003	OS	TRT1	26.1	2
1004	OD	TRT1	20.3	1
1004	OS	TRT1	26.2	2

TABLE 5.2

The Annotated Dataset to Display the Mean Values by Treatment and Eye

FUNCTION	COLOR	XSYS	YSYS	HSYS	X	Y	SIZE
MOVE	RED	2	2	4	0.8	24.69	2
DRAW	RED	2	2	4	1.2	24.69	2
MOVE	RED	2	2	4	1.8	25.03	2
DRW	RED	2	2	4	2.2	25.03	2
MOVE	RED	2	2	4	2.8	25.03	2
DRAW	RED	2	2	4	3.2	25.03	2
MOVE	RED	2	2	4	3.8	25.8	2
DRW	RED	2	2	4	4.2	25.8	2

group using the *LINE* function. The labels for the subject numbers for each treatment group are included in the same SG annotation dataset. The SG annotated dataset is shown in Table 5.3. More detailed instructions on SG annotation can be found in Chapters 10 and 11 of the *SAS 9.3 ODS Graphics: Procedures Guide* (SAS Institute Inc., 2012).

5.4.2 Notes to SAS Programs

Detailed SAS programs to produce the sample plots using both the GPLOT and the SGPLOT procedures are provided in the Appendix (Section 5.7). The programs consist of the following main sections and points.

5.4.2.1 Data Simulation

- IOP data for 200 subjects are simulated for both eyes (OD and OS) from a normal distribution with a mean of 25 mm Hg and a SD

TABLE 5.3

SG Annotated Dataset to Display the Mean Values by Treatment and Eye and Labels with Subject Numbers of Treatments

DRAWSPACE	function	x1	y1	x2	y2	LINECOLOR	anchor	label	textcolor
datavalue	line	0.8	24.69	1.2	24.69	Red			
datavalue	line	1.8	25.03	2.2	25.03	Red			
datavalue	line	2.8	25.03	3.2	25.03	Red			
datavalue	line	3.8	25.8	4.2	25.8	Red			
datavalue	text	0.5	35	4.2	25.8	Red	TOPLEFT	TRT1(N = 118)	green
datavalue	text	2.5	35	4.2	25.8	Red	TOPLEFT	TRT2(N = 82)	green

of 3.5. Subjects are randomly assigned to groups TRT1 or TRT2 at a 1:1 ratio.

- An *index* variable is generated to represent the 2 treatments and 2 eyes, and will be used for plotting on the x-axis.

5.4.2.2 Data Analyses and Manipulation

- The subject number of each treatment group is determined by using *PROC FREQ* and saved into macro variables (*num_trt1, num_trt2*) by using the call symputx() function.
- A macro *LineUp_Jitter (idn =, xvar =, yvar =, lvl =, jitter =, odn =)* is created by using the algorithms specified in Section 5.2.
  - The minimum and the maximum values in a dataset are determined using *Proc Means* and the values are saved in macro variables (Min and Max).
  - Since the minimum and the maximum values of the data might not be an integer number, the function *floor(&min.)* is used to return the largest integer that is less than or equal to the minimum value; and the function *ceil(&max.)* is used to return the smallest integer that is greater than or equal to the maximum value. The data points are classified into different line-up groups (intervals) starting from *floor(&min.)* to *ceil(&max.)*. The data points positioned in the same line-up group (interval) can differ from each other by the &lvl. unit.

```
do i = floor(&min.) to ceil(&max.) by &lvl.;
   if i < = &yvar. < i+1 then yvar_int = i;
end;
```

  - The number of data points at each line-up group (interval) are calculated using *Proc Freq* and saved in a dataset *freqdata*.
  - The data points at each group (interval) are jittered based on the number of data points within the group using the different algorithms as specified in Section 5.2.

```
if count = 1 then do;
   xvar_j = &xvar.;
   output;
end;
if count>1 then do;
   if mod(count,2) > 0 then do; ** odd count number;
      xvar_j = &xvar.; pos = &xvar.; neg = &xvar.; output;
      do i = 1 to count-1 by 1;
```

```
         if pos_neg = 1 then do;
           xvar_j = neg-&jitter;
           neg = xvar_j;
         end;
         else do;
           xvar_j = pos+&jitter;
           pos = xvar_j;
         end;
         output;
         pos_neg = -1*pos_neg;
       end;
     end;
     if mod(count,2) = 0 then do; ** even count number;
       pos = &xvar.; neg = &xvar.;
       do i = 1 to count by 1;
         if pos_neg = 1 then do;
           xvar_j = neg-&jitter/2;
           neg = xvar_j-&jitter/2;
         end;
         else do;
           xvar_j = pos+&jitter/2;
           pos = xvar_j+&jitter/2;
         end;
         output;
         pos_neg = -1*pos_neg;
       end;
     end;
   end;
 end;
run;
```

- The original variables (&yvar. &xvar.) and the jittered and the assigned interval variables (xvar_j yvar_int) are kept in the output dataset (*&odn*).

- The macro *LineUp_Jitter ()* is invoked to generate a line-up jittered dataset, with the jittering level of 0.08 on the x-axis, and the line-up level of 0.5 mm Hg on the y-axis. Both the original IOP and the reassigned IOP values are saved in the line-up jittered dataset (IOP_Jittered).

- The mean IOP value for each eye of each treatment group is calculated using *PROC MEANS* and saved in a dataset with the mean value for each eye and treatment group saved to macro variables (*MN_Index1* to *MN_Index4*).

5.4.2.3 Producing Sample Figures Using PROC GPLOT

- An annotated dataset (*IOP_ANNOT*) is created to draw short lines to represent the mean values for each eye of each treatment. The SAS annotate macros *%ANNOMAC,%DCLANNO,* and *%LINE* are used to create annotate datasets. Please see Chapter 3 for the descriptions of the three SAS built-in macros.

- Using the line-up jittered dataset (*IOP_JITTERED*), two line-up jittered scatter plots (Figures 5.1 and 5.2) are produced, with Figure 5.1 using the original IOP values (with the variable IOP) and Figure 5.2 using the reassigned IOP values (with the variable yvar_int) on the y-axis.

- Four SYMBOL statements are used to specify the symbol, font, and color for the scatters for the 2 eyes in the 2 treatment groups. Note that albany amt/Unicode is used in the FONT option with the value of '25cb'x representing open circles and 25cf'x representing the closed circles. These are the font-based markers and provide nicer and smoother (anti-aliased) markers than the symbol markers (circle and dot). A list of Unicodes for commonly used geometric shapes can be found at the link http://www.unicode.org/charts/PDF/U25A0.pdf.

```
SYMBOL1 H = 3 C = BLACK    CO = BLACK I = NONE font =
'albany amt/unicode' VALUE = '25cb'x;
SYMBOL2 H = 3 C = BLUE     CO = BLUE I = NONE font =
'albany amt/unicode' VALUE = '25cf'x;
SYMBOL3 H = 3 C = BLACK    CO = BLACK I = NONE font =
'albany amt/unicode' VALUE = '25cb'x;
SYMBOL4 H = 3 C = BLUE     CO = BLUE I = NONE font =
'albany amt/unicode' VALUE = '25cf'x;

AXIS1 OFFSET = (1,1) ORDER = (10 to 35 by 5) LABEL =
(FONT = "&FONTNAME" h = 2.5 ANGLE = 90 "IOP Value (mm Hg)")
VALUE = (H = 2.5) MINOR = NONE;
AXIS2 MAJOR = NONE MINOR = NONE ORDER = (0.5 TO 4.5 BY 0.5)
OFFSET = (1,1)
  LABEL = (FONT = "&FONTNAME" h = 2.5 "Treatment Group
and Eye")
  reflabel = (position = top c = green h = 2.5 j = r
FONT = "&FONTNAME." " TRT1(N = &num_trt1.)" " TRT2(N =
&num_trt2.)");

PROC GPLOT DATA = iop_jittered;
  FILENAME GSASFILE "&OUTLOC./Figure 5.2.&EXT.";
  PLOT iop * xvar_j = index/HAXIS = AXIS2 VAXIS = AXIS1
noframe nolegend
    href = 0.5, href = 2.5 ANNOTATE = IOP_ANNOT;
    FORMAT xvar_j indexdf.;
RUN;
```

```
PLOT yvar_int * xvar_j = index/HAXIS = AXIS2 VAXIS = AXIS1
noframe nolegend
  href = 0.5, href = 2.5 ANNOTATE = IOP_ANNOT;
  FORMAT xvar_j indexdf.;
RUN;
quit;
```

5.4.2.4 Producing Sample Figures Using PROC SGPLOT

- An SG annotate dataset is created to display the reference line labels for treatment group and subject numbers and to draw short lines to indicate the mean values for each eye at each treatment.

- The attribute map option in SGPLOT is used to display the attributes (marker symbols and marker color) for the grouping variable—the index (the 4 combinations of the treatment groups and eyes). The attribute map is a new feature in SAS 9.3, which is used to define the visual attributes of particular group of values (Mantange and Heath, 2011).

```
** Using the attribute map to indicate attributes and
colors for the groups
data symbol_attrs;
  retain id "Indexid";
  length value 5 markercolor $ 6 markersymbol $ 15;
  input value markercolor $ markersymbol $;
  cards;
  1 black circle
  2 blue circlefilled
  3 black circle
  4 blue circlefilled
  ;
run;

proc sgplot data = iop_jittered sganno = anno_sg dattrmap
= symbol_attrs noautolegend;
  scatter x = xvar_j y = iop / group = index attrid = indexid;
  xaxis VALUES = (0.5 to 4.5 by.5) label = "Treatment
Group and Eye";
  yaxis VALUES = (10 to 35 by 5) label = "IOP Value (mm Hg)";
  REFLINE 0.5 2.5/AXIS = x;
  format xvar_j indexdf.;
run;
proc sgplot data = iop_jittered sganno = anno_sg dattrmap
= symbol_attrs noautolegend;
  scatter x = xvar_j y = yvar_int / group = index attrid
= indexid;
```

```
   xaxis VALUES = (0.5 to 4.5 by.5) label = "Treatment Group
and Eye";
   yaxis VALUES = (10 to 35 by 5) label = "IOP Value (mm Hg)";
   REFLINE 0.5 2.5/AXIS = x;
   format xvar_j indexdf.;
run;
```

- Another way to produce scatter plots for a group (eye and treatment) without using the attribute map feature is to use 4 scatter statements with 4 different variables for the y-axis.

```
proc sgplot data = iop_fig sganno = anno_sg noautolegend;
   scatter x = xvar_j y = jiop1/markerattrs = (symbol =
circle color = black size = 10);
   scatter x = xvar_j y = jiop2/markerattrs = (symbol =
circlefilled color = blue size = 10);
   scatter x = xvar_j y = jiop3/markerattrs = (symbol =
circle color = black size = 10);
   scatter x = xvar_j y = jiop4/markerattrs = (symbol =
circlefilled color = blue size = 10);
   xaxis VALUES = (0.5 to 4.5 by.5) label = "Treatment
Group and Eye";
   yaxis VALUES = (10 to 35 by 5) label = "IOP Value (mm Hg)";
   REFLINE 0.5 2.5/AXIS = x;
   format xvar_j indexdf.;
run;
```

5.5 Summary and Discussion

The key to producing a line-up jittered scatter plot is to jitter the data points based on the number of data at each line-up group and prepare the line-up jittered dataset for plotting. Detailed algorithms for line-up jittering are discussed and an SAS macro for line-up jittering is developed that can be modified and used to produce your own customized line-up jittered scatter plots.

The attribute map is a new feature beginning in SAS 9.3, and is used to define the visual attributes of particular group values (Matange and Heath, 2011). The *DATTRMAP* option is used to specify which group *VALUE* should get which marker symbol or color. In SAS 9.2, the attribute map feature is not available, and we will need to define a new style to replace the symbol and color values for *GraphData1-N* with the values that you want (Heath, 2012).

An easy way to do this is to use the *MODSTYLE* macro (Matange, 2012). In SAS 9.4, there is a way to set group colors, symbols, and line patterns in the procedure syntax itself.

The other differences in using the GLOT and SGPLOT procedures in producing the scatter plots were discussed in Chapter 4.

5.6 References

Heath, D. 2012. "Roses are Red, Violets are Blue …" SAS blogs, February 27, http://blogs.sas.com/content/graphicallyspeaking/2012/02/27/roses-are-red-violets-are-blue/.

Matange, S. 2012. "Quick and Easy with MODSTYLE." SAS blogs, June 14, http://blogs.sas.com/content/graphicallyspeaking/2012/06/14/quick-and-easy-with-modstyle/.

Matange, S., and Heath, D. 2011. *Statistical Graphics Procedures by Example: Effective Graphs Using SAS®.* Cary, NC: SAS Institute Inc.

SAS Institute Inc. 2012. *SAS/GRAPH® 9.3: Reference,* 3rd ed. Cary, NC: SAS Institute Inc.

SAS Institute Inc. 2012. *SAS® 9.3 ODS Graphics: Procedures Guide,* 3rd ed. Cary, NC: SAS Institute Inc.

5.7 Appendix: SAS Programs for Producing the Sample Figures

```
**************************************************************;
* Program Name: Line-Up Jittered Scatter Plots.sas        *;
* The program produces the below line-up jittered scatter *;
* graphs                                                  *;
* Figure 5.1 Line-up Jittered Scatter Plot for IOP by     *;
* Treatment and Eye                                       *;
* - IOP on the Y-axis are the original values             *;
* Figure 5.2 Line-up Jittered Scatter Plot for IOP by     *;
* Treatment and Eye                                       *;
* - IOP Values within the same line on the Y-axis might   *;
* differ from each other by 0.5 mmHg                      *;
* Author: Charlie Liu                                     *;
**************************************************************;
options mprint symbolgen nodate nonumber validvarname = v7
orientation = landscape;
%let pgmname = Line-up Jittered Scatter Plots.sas;
%let outloc = C:\SASBook\Sample Figures\Chapter 5;
%let pgmloc = C:\SASBook\SAS Programs;
```

```
%let pgmpth = &pgmloc.\&pgmname. &sysdate9. &systime. SAS
V&sysver.;

** setting-up macros for data simulation;
%let seed = 05;%let subjnum = 200;%let mean = 25;%let sd 3.5;
proc format;
  value trtdf
    1 = 'TRT1'
    2 = 'TRT2'
    OTHER = ' ';
  value eyedf
    1 = 'OD'
    2 = 'OS'
    OTHER = ' ';
  value indexdf
    1, 3 = 'OD'
    2, 4 = 'OS'
    OTHER = ' ';
run;

** Simulate subject's IOP data in both eyes;
data iop;
  do i = 1 to &subjnum.; ** num. of subjects;
    if ranuni (&seed.) < = 0.5 then trt = 1;
    else trt = 2;
    do j = 1 to 2; ** 2 eyes;
      subjid = 1000 + i;
      eye = j;
      iop = round((RANNOR(&seed. + i + j)* &sd. + &mean.),.1);
      output;
    end;
  end;
  format eye eyedf. trt trtdf.;
  drop i j;
run;

** Create an index variable to combine the treatment group and
eye;
data iop;
  set iop;
  if trt = 1 and eye = 1 then index = 1;
  if trt = 1 and eye = 2 then index = 2;
  if trt = 2 and eye = 1 then index = 3;
  if trt = 2 and eye = 2 then index = 4;
run;

** Number of subjects by treatment;
proc sort data = iop out = subj_trt nodupkey;
  by trt subjid;
run;
```

```
proc freq data = subj_trt noprint;
  table trt/out = trt_freq (drop = percent);
run;

** save the subject number at each treatment group to macro
variables;
data _null_;
  set trt_freq;
  call symputx('num_Trt'||left(put(trt,best.)), put(count, 3.0));
run;

**************************************************************;
* Macro name: LineUp_Jitter(idn =, xvar =, yvar =, lvl =,    *;
* jitter =, odn =)                                           *;
* Input variables:                                           *;
* idn = input dataset name, xvar = variable for x-axis,      *;
* yvar = variable for y-axis, lvl = interval level for       *;
* y-axis, jitter = jitter level, odn = output dataset name   *;
**************************************************************;

%macro LineUp_Jitter (idn =, xvar =, yvar =, lvl =, jitter =,
odn =);
** Get the Min and Max values and save them into macro
variables;
PROC MEANS DATA = &idn. MIN MAX maxdec = 2 noprint;
  VAR &yvar.;
  OUTPUT OUT = tmp_STAT min = min max = max;
RUN;
data _null_;
  set tmp_stat;
  call symput('Min', min);
  call symput('Max', max);
run;

proc sort data = &idn. out = d_tmp;
  by &yvar.;
run;

data add_tmp;
  set d_tmp;
  do i = floor(&min.) to ceil(&max.) by &lvl.;
    if i < = &yvar. < i+1 then yvar_int = i;
  end;
run;

** get the number of data points at each interval by x-axis
value;
proc freq data = add_tmp noprint;
  tables &xvar.*yvar_int/out = freqdata;
run;
```

```
data jitter_sum;
   retain pos_neg 1;
   set freqdata;
   pos_neg = 1;

   if count = 1 then do;
      xvar_j = &xvar.;
      output;
   end;

   if count>1 then do;
      if mod(count,2) > 0 then do; ** odd count number;
         xvar_j = &xvar.; pos = &xvar.; neg = &xvar.; output;
         do i = 1 to count-1 by 1;
            if pos_neg = 1 then do;
               xvar_j = neg-&jitter;
                  neg = xvar_j;
            end;
            else do;
               xvar_j = pos+&jitter;
                  pos = xvar_j;
            end;
            output;
            pos_neg = -1*pos_neg;
         end;
      end;

      if mod(count,2) = 0 then do; ** even count number;
         pos = &xvar.; neg = &xvar.;
         do i = 1 to count by 1;
            if pos_neg = 1 then do;
               xvar_j = neg-&jitter/2;
                  neg = xvar_j-&jitter/2;
            end;
            else do;
               xvar_j = pos+&jitter/2;
                  pos = xvar_j+&jitter/2;
            end;
            output;
            pos_neg = -1*pos_neg;
         end;
      end;
   end;
run;

proc sort data = add_tmp;
   by yvar_int &xvar.;
run;
```

```
proc sort data =

jitter_sum;
   by yvar_int &xvar.;
run;

data jitter_sum;
   set jitter_sum;
   by yvar_int &xvar.;
   ord = _n_;
   run;

data add_tmp;
   set add_tmp;
   by yvar_int &xvar.;
   ord = _n_;
   run;

** Output dataset: Jittered dataset for ploting;
data &odn.;
   merge jitter_sum (in = a) add_tmp;
   by ord;
   if a;
   keep &yvar. &xvar. yvar_int xvar_j;
%mend LineUp_Jitter;

** Producing line-up jittered datasets;
** Jitter Level on X-axis = 0.08, Line-up Interval on y-axis =
0.5 mmHg;
%LineUp_Jitter (idn = IOP, yvar = iop, xvar = index, lvl = 0.5,
jitter = 0.08, odn = IOP_Jittered);

* Summary Statistics by index (i.e. each eye at each treatment);
PROC MEANS DATA = IOP NWAY MEAN STD STDERR MIN MAX MEDIAN
maxdec = 2 noprint;
   CLASS index;
   VAR iop;
   OUTPUT OUT = iop_STAT MEAN = Mean;
RUN;

** save the mean value at each treatment group and eye (index)
to macro variables;
data _null_;
   set iop_STAT;
   call symputx('MN_Index'||left(put(index,best.)),
put(mean, 5.2));
run;

* Make the SAS Annotate data set macros available for use.;
%ANNOMAC;
```

```
* Generate an annotation data set that draw indicators for
means;
DATA IOP_ANNOT;
  SET IOP_STAT;
  %DCLANNO;
  RESULT_MN = round(mean,.01);
  RETAIN SEMULT 1 NUM_OFFSET.1 SHIFT_VAL.20;
  SIZE = 2;
  HSYS = '4';
  XSYS = '2';
  YSYS = '2';
  %* draw the horizontal line for mean *;
  %LINE(index-SHIFT_VAL, RESULT_MN, index+SHIFT_VAL, RESULT_MN,
RED, 1, SIZE);
  KEEP X Y FUNCTION COLOR SIZE HSYS XSYS YSYS;
RUN;

%LET FONTNAME = Times;
%LET DRIVER = PSCOLOR;%LEt EXT = PS;
goptions
  reset    = all
  GUNIT    = PCT
  rotate   = landscape
  gsfmode  = replace
  gsfname  = GSASFILE
  device   = &DRIVER
  hsize    = 8 in
  horigin  = 0 in
  vsize    = 6.5 in
  vorigin  = 0 in
  ftext    = "&FONTNAME"
  htext    = 10pt
  ftitle   = "&FONTNAME"
  htitle   = 10pt
;

SYMBOL1 H = 3 C = BLACK    CO = BLACK I = NONE font = 'albany
amt/unicode' VALUE = '25cb'x;
SYMBOL2 H = 3 C = BLUE     CO = BLUE I = NONE font = 'albany
amt/unicode' VALUE = '25cf'x;
SYMBOL3 H = 3 C = BLACK    CO = BLACK I = NONE font = 'albany
amt/unicode' VALUE = '25cb'x;
SYMBOL4 H = 3 C = BLUE     CO = BLUE I = NONE font = 'albany
amt/unicode' VALUE = '25cf'x;

AXIS1 OFFSET = (1,1) ORDER = (10 to 35 by 5) LABEL = (FONT =
"&FONTNAME" h = 2.5 ANGLE = 90 "IOP Value (mm Hg)") VALUE =
(H = 2.5) MINOR = NONE;
AXIS2 MAJOR = NONE MINOR = NONE ORDER = (0.5 TO 4.5 BY 0.5)
OFFSET = (1,1)
  LABEL = (FONT = "&FONTNAME" h = 2.5 "Treatment Group and Eye")
  reflabel = (position = top c = green h = 2.5 j = r FONT =
"&FONTNAME." " TRT1(N = &num_trt1.)" " TRT2(N = &num_trt2.)");
```

```
title1 "Figure 5.1 Line-up Jittered Scatter Plot for IOP by
Treatment and Eye";
title2 "- IOP on the Y-axis are the original values";
footnote1 "Note: Red line denotes the mean value within each
treatment and eye";
footnote2 "&pgmpth.";
ods listing;
PROC GPLOT DATA = iop_jittered;
   FILENAME GSASFILE "&OUTLOC./Figure 5.1.&EXT.";
   PLOT iop * xvar_j = index/HAXIS = AXIS2 VAXIS = AXIS1 noframe
nolegend
      href = 0.5, href = 2.5 ANNOTATE = IOP_ANNOT;
      FORMAT xvar_j indexdf.;
   RUN;

   title1 "Figure 5.2 Line-up Jittered Scatter Plot for IOP by
Treatment and Eye";
   title2 "- IOP Values within the same line on the Y-axis
might differ from each other by 0.5 mmHg";
   footnote1 "Note: Red line denotes the mean value within each
treatment and eye";
   footnote2 "&pgmpth.";
   FILENAME GSASFILE "&OUTLOC./Figure 5.2.&EXT.";
   PLOT yvar_int * xvar_j = index/HAXIS = AXIS2 VAXIS = AXIS1
noframe nolegend
      href = 0.5, href = 2.5 ANNOTATE = IOP_ANNOT;
      FORMAT xvar_j indexdf.;
   RUN;
quit;

*****************************************************************;
** Reproduce the same figures using the SGPLOT procedure     *;
*****************************************************************;
** SG Annotatin dataset to display reference labels and mean
values of each treatment and eye;
data anno_sg;
   retain DRAWSPACE "datavalue";
   ** Using Function LINE to draw the mean lines from (x1, y1)
to (x2, y2);
   function = "line"; x1 = 0.8; y1 = &MN_Index1.; x2 = 1.2;
y2 = &MN_Index1.;
      LINECOLOR = "Red"; output;
   function = "line"; x1 = 1.8; y1 = &MN_Index2.; x2 = 2.2;
y2 = &MN_Index2.;
      LINECOLOR = "Red"; output;
   function = "line"; x1 = 2.8; y1 = &MN_Index3.; x2 = 3.2;
y2 = &MN_Index3.;
      LINECOLOR = "Red"; output;
   function = "line"; x1 = 3.8; y1 = &MN_Index4.; x2 = 4.2;
y2 = &MN_Index4.;
      LINECOLOR = "Red"; output;
```

```
   ** Using Function TEXT to label the subject numbers in the
2 TRT groups at (x1, y1) and (x2, y2);
   function = "text"; x1 = 0.5; y1 = 35; anchor = "TOPLEFT";
label = "TRT1(N = &num_trt1.)";
      textcolor = "green"; output;
   function = "text"; x1 = 2.5; y1 = 35; anchor = "TOPLEFT";
label = "TRT2(N = &num_trt2.)";
      textcolor = "green"; output;
run;

** Using the attribute map to indicate attributes and colors
for the grouping in SGPLOT;
data symbol_attrs;
   retain id "Indexid";
   length value 5 markercolor $ 6 markersymbol $ 15;
   input value markercolor $ markersymbol $;
   cards;
   1 black circle
   2 blue circlefilled
   3 black circle
   4 blue circlefilled
   ;
run;

%LET OUTPUTFMT = PS;
%LET FMTTITLE = H = 1.5 JUSTIFY = CENTER FONT = "&FONTNAME";
%LET FMTFOOT = H = 1 JUSTIFY = CENTER FONT = "&FONTNAME";
ods listing gpath = "&outloc.";

title1 "Figure 5.1 Line-up Jittered Scatter Plot for IOP by
Treatment and Eye";
title2 "IOP on the Y-axis are the original values";
footnote1 "Note: Red line denotes the mean value within each
treatment and eye";
footnote2 "&pgmpth.";
ods graphics/reset = all width = 8in height = 6in noborder
OUTPUTFMT = &OUTPUTFMT. imagename = "FigSG 5_1";
proc sgplot data = iop_jittered sganno = anno_sg dattrmap =
symbol_attrs noautolegend;
   scatter x = xvar_j y = iop / group = index attrid = indexid;
   xaxis VALUES = (0.5 to 4.5 by.5) label = "Treatment Group
and Eye";
   yaxis VALUES = (10 to 35 by 5) label = "IOP Value (mm Hg)";
   REFLINE 0.5 2.5/AXIS = x;
   format xvar_j indexdf.;
run;

title1 "Figure 5.2 Line-up Jittered Scatter Plot for IOP by
Treatment and Eye";
title2 "IOP Values within the same line on the Y-axis might
differ from each other by 0.5 mmHg";
```

```
footnote1 "Note: Red line denotes the mean value within each
treatment and eye";
footnote2 "&pgmpth.";
ods graphics/reset = all width = 8in height = 6in noborder
OUTPUTFMT = &OUTPUTFMT. imagename = "FigSG 5_2";
proc sgplot data = iop_jittered sganno = anno_sg dattrmap =
symbol_attrs noautolegend;
  scatter x = xvar_j y = yvar_int / group = index attrid =
indexid;
  xaxis VALUES = (0.5 to 4.5 by.5) label = "Treatment Group
and Eye";
  yaxis VALUES = (10 to 35 by 5) label = "IOP Value (mm Hg)";
  REFLINE 0.5 2.5/AXIS = x;
  format xvar_j indexdf.;
run;

** Reproduce Figure 5.2 sgplot without using the attribute map;
data iop_fig;
  set iop_jittered;
  if index = 1 then jiop1 = yvar_int;
  if index = 2 then jiop2 = yvar_int;
  if index = 3 then jiop3 = yvar_int;
  if index = 4 then jiop4 = yvar_int;
run;

ods listing gpath = "&outloc.";
ods graphics/reset = all width = 8in height = 6in noborder
OUTPUTFMT = emf imagename = "Fig 5_2b";

title1 "Figure 5.2 Line-up Jittered Scatter Plot for IOP by
Treatment and Eye";
title2 "IOP Values within the same line on the Y-axis might
differ from each other by 0.5 mmHg";
footnote1 "Note: Red line denotes the mean value within each
treatment and eye";
footnote2 "&pgmpth.";
proc sgplot data = iop_fig sganno = anno_sg noautolegend;
  scatter x = xvar_j y = jiop1 / markerattrs = (symbol =
circle color = black size = 10);
  scatter x = xvar_j y = jiop2 / markerattrs = (symbol =
circlefilled color = blue size = 10);
  scatter x = xvar_j y = jiop3 / markerattrs = (symbol =
circle color = black size = 10);
  scatter x = xvar_j y = jiop4 / markerattrs = (symbol =
circlefilled color = blue size = 10);

  xaxis VALUES = (0.5 to 4.5 by.5) label = "Treatment Group
and Eye";
  yaxis VALUES = (10 to 35 by 5) label = "IOP Value (mm Hg)";
  REFLINE 0.5 2.5/AXIS = x;
  format xvar_j indexdf.;
run;
```

6

Thunderstorm or Raindrop Scatter Plots

6.1 Introduction

Data visualization, using attractive and effective figures to display all data points to allow readers to see the overall data patterns, study the inherent relationships, and detect outliers has been an interesting yet challenging job for statisticians (Cleveland, 1985; Tufte, 1983, 1997, 2006). Many different types of plots have been introduced and produced, including those especially named for their shapes, like the bag, bar, bike, box, bubble, pie, pyramid, spaghetti, rainbow, violin, and waterfall plots (Allison, 2012; Hyndman and Shang, 2010). This chapter introduces and illustrates how to produce a new type of plot known for its shape, the thunderstorm or raindrop scatter plot. This plot allows the viewing of data with two or more values on the y-axis corresponding to one value on the x-axis for each of several subjects in a population. The resulting plot looks like raindrops, with each raindrop representing data for a single subject. When data for many subjects are plotted, it resembles a thunderstorm, hence the name. A thunderstorm/raindrop scatter plot is a useful tool for data visualization and outlier detection. The concept of thunderstorm/raindrop scatter plots is introduced and sample figures are produced using the SAS/GRAPH annotate facility in PROC GPLOT and the HIGHLOW statement in PROC SGPLOT.

The author presented part of the contents in this chapter at the SAS Global Forum, 2013 (Liu, 2013).

6.2 Application Examples

Thunderstorm or raindrop scatter plots have applications in many areas, including clinical research, agriculture, and finance. For example, these plots can be used to display efficacy measurement before and after treatment for patients in clinical trials, yield production before and after fertilizer application for crop varieties in agriculture, and buying power before and after a certain promotion for customers in a financial institutions.

To illustrate the application and production of thunderstorm/raindrop scatter plots, three examples are presented. The first two are based on clinical research in the glaucoma therapeutic area, and the third in agriculture. The first example is for two types of intraocular pressure (IOP) values on the y-axis corresponding to central corneal thickness (CCT) on the x-axis for each eye of 100 patients. The second example is for mean diurnal IOP at baseline and Week 12 on the y-axis by subjects grouped within investigator sites on the x-axis. The third example is for plant nutrient contents at three salinity levels on the y-axis by the plant mineral nutrients on the x-axis. The data structure and SAS annotated dataset used to produce the sample plots are discussed. The SAS programs that produced the sample figures are discussed and included in the Appendix (Section 6.6).

6.2.1 Example 1: Two IOPs on the y-Axis by CCT on the x-Axis

The Goldmann applanation tonometer (GAT) is the gold standard for measuring IOP, and has been the most widely used IOP measurement technology (Goldmann and Schmidt, 1957). However, GAT IOP has been known to be affected by eye physical ocular properties, especially the CCT (Ehlers et al., 1975; Whitacre and Stein, 1993). Recently, the Pascal dynamic contour tonometer (DCT, Swiss Microtechnology AG, Port, Switzerland) was developed to provide automated IOP readings that are less affected by CCT compared with those measured by GAT (Boehm, et al. 2008). A thunderstorm/raindrop scatter plot is a good tool to study the agreement/difference between GAT and DCT IOPs and see whether they are affected by the corresponding CCT values.

We simulated a dataset including 100 patients, with IOP values by GAT and DCT and the corresponding CCT value for each eye (see data structure in Table 6.2). The thunderstorm scatter plot displaying the two IOPs (GAT and DCT) on the y-axis by its corresponding CCT value on the x-axis at the eye level of all 100 patients (200 eye data points) is shown in Figure 6.1. The figure allows us to view the two IOP values (GAT and DCT) of the same eye and directly see its agreement or difference by CCT values. If you feel the thunderstorm plot is too crowded with too many data points (raindrops), you can separate it into several figures by subsetting the data by patients (e.g., with each figure displaying data from 20 eyes in 10 patients) or by the CCT values (500 to 510, 520 to 540 microns, etc.). The new figures might not be qualified as a "thunderstorm" anymore, but as "raindrop" plots (Figure 6.2 displays data from 20 eyes of 10 patients).

6.2.2 Example 2: Baseline and Week 12 Mean Diurnal IOP on the y-Axis by Subjects Grouped by Investigator Sites on the x-Axis

Let's design a virtual clinical trial. Subjects' IOP values are measured at hours 0, 2, and 8 at baseline (day 0), and at Weeks 4, 8, and 12 postbaseline. Mean diurnal IOP at week 12, the mean IOP values at hours 0, 2, and 8 at Week 12 in the study eye, is the primary analysis variable.

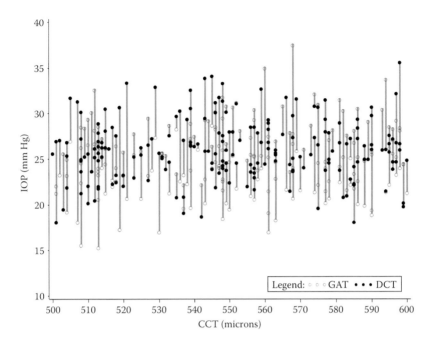

FIGURE 6.1
Thunderstorm scatter plot for GAT and DCT IOP by CCT value for 200 eyes in 100 patients.

The thunderstorm scatter plot in Figure 6.3 provides a good visualization of the mean diurnal IOP values at baseline and week 12 for subjects within each investigator site. The plot allows us to see the overall data pattern for all subjects at both visits within each investigator site and helps the clinical team to issue data queries. For example, there are two subjects (one at investigator site 5 and the other at site 9) whose IOP values at week 12 are higher than at baseline. These values are considered as outliers because the subjects' IOP values are expected to decrease after treatment following baseline.

6.2.3 Example 3: Plant Tissue Nutrient Contents before and after Salinity Stress in Creeping Bentgrass

Creeping bentgrass (*Agrostis palustris*) is widely used in golf course putting greens, especially in the northern part of the United States, for its excellent tolerance to low mowing and its smooth putting surface. However, in some golf courses near coastal areas, the grass is constantly under salinity stress due to seawater invasion. It would be interesting to understand the effect of salinity on the plant growth, especially on the essential mineral nutrient absorption and accumulation. Researchers at North Carolina State University studied the influence of three levels of salinity (none, medium, and high) on the plant tissue nutrient contents for creeping bentgrass using a controlled greenhouse hydroponic system (Liu, 1998).

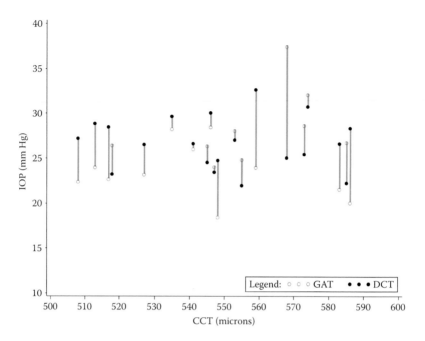

FIGURE 6.2
Raindrop scatter plot for GAT and DCT IOP by CCT value for 20 eyes in 10 patients.

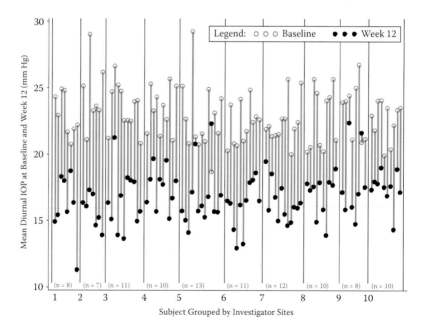

FIGURE 6.3
Thunderstorm scatter plot for study of eye mean diurnal IOP at baseline and Week 12.

TABLE 6.1

Plant Tissue Nutrient Content before and after Salinity Stress
in Creeping Bentgrass

Salinity Stress (ds/m, EC)	N	P	K	Ca	Mg	S	Cl	Na
0	60.4	6.24	37.9	6.25	3.00	5.35	7.26	0.69
8	58.2	5.69	26.5	3.4	3.38	5.01	16.9	8.39
16	56.7	5.04	24.2	3.41	3.61	4.99	20.9	9.23

Source: Data from C. Liu, "Effects of Humic Substances on Creeping
Bentgrass Growth and Stress Tolerance." PhD diss., North Carolina
State University (1998).

The original results are displayed in Table 6.1. Using the data, a raindrop
scatter plot was generated to position the plant nutrient names on the x-axis
and the plant tissue nutrient content (g/kg) at the three salinity levels (0, 8,
and 16 ds/m EC) on the y-axis (Figure 6.4). The nutrient content of the same
element at the three salinity levels was connected together to form a raindrop. Without doubt, the raindrop scatter plot provides a better data visualization and summary of the results than the table. We can easily conclude
that the salinity had little or no effect on the absorption and accumulation
of phosphorus (P), magnesium (Mg), and sulfur (S), slightly decreased the

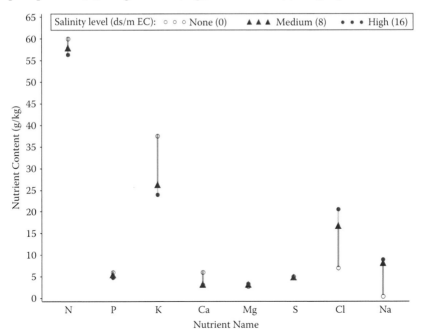

FIGURE 6.4
Raindrop scatter plot for plant nutrient contents at three salinity levels.

absorption of nitrogen (N) and calcium (Ca), significantly decreased the absorption of potassium (K), and increased the absorption of chloride (Cl) and sodium (Na).

6.3 Producing the Sample Figures

6.3.1 Data Structure and SAS Annotated Dataset

Part of the simulated dataset with 100 subjects' IOP and CCT values at the eye level used to produce the thunderstorm/rain-drop scatter plots in Figure 6.1 is shown in Table 6.2. Each subject's eye (right eye [OD] and left eye [OS]) has IOP values by GAT and DCT and the corresponding CCT value. Each subject has 4 recorded values in the dataset (each eye with 2 different IOPs) and one corresponding CCT value. There are 100 subjects with a total of 400 records in the whole simulated dataset.

The SAS/GRAPH annotate facility is used to connect the 2 IOPs of the same eye together with a straight line. The record for one subject in the annotated dataset is displayed in Table 6.3. For the right eye (OD), SAS at first moves the point to the position at (585, 26.9) (CCT and GAT IOP values on the x- and y-axis, respectively), then draws a line to the point (585, 20.9) in vivid green color (VILG). The line type and size is controlled by the value of the variables *line* and *size*. For a more detailed introduction to the SAS annotate facility, please refer to Section 5 of the "SAS® 9.3 Product Documentation" (SAS Institute Inc., 2013).

6.3.2 Notes to SAS Programs

The three SAS programs used to produce the four sample figures using both the GPLOT and the SGPLOT procedures are provided in the Appendix (Section 6.6).

TABLE 6.2

Part of the Simulated Dataset to Produce Figure 6.1

SUBJID	EYE	IOP_TYPE	IOP	CCT
1001	OD	GAT	26.9	585
1001	OD	DCT	20.9	585
1001	OS	GAT	24.1	559
1001	OS	DCT	31.3	559

TABLE 6.3

Part of the Annotated Dataset Used to Connect Data Points to Form Raindrops in Figure 6.1

SUBJID	EYE	CCT	IOP_GAT	IOP_DCT	RECID	function	xsys	ysys	x	y	color	size	line
1001	OD	585	26.9	20.9	1	move	2	2	585	26.9	—	—	—
1001	OD	585	26.9	20.9	1	draw	2	2	585	20.9	VILG	2	1
1001	OS	559	24.1	31.3	2	move	2	2	559	24.1	—	—	—
1001	OS	559	24.1	31.3	2	draw	2	2	559	31.3	VILG	2	1

6.3.2.1 Main Sections and Features of the First Program

6.3.2.1.1 Dataset Simulation

- A virtual clinical trial is simulated to include 100 subjects. Each eye of a subject has two IOPs (GAT and DCT) and one CCT. The IOP and CCT values are assigned based on normal distributions with the preset mean and standard deviation (SD).

- In simulating the SAS dataset, DO loops and SAS function *RANNOR* () are used to generate random numbers that are normally distributed with preset sample means and standard deviations. The fixed seed number is used so that each independent run will produce the same datasets and the same figures. It is also a good practice to use a different seed at each DO loop, which is achieved by incorporating the loop numbers into the seed by using "(&seed. + i + j)" as the seed at each loop.

6.3.2.1.2 Producing Figure 6.1 Using the GPLOT Procedure

- An SAS annotate facility is used to draw a line between 2 points to from a raindrop in *PROC GPLOT* using the *anno =* option. This line is drawn by "moving" and "drawing" a line between 2 points. Please note that the *RecID* variable in the annotated dataset is based on the eye level and each subject has 2 records. This fact is important in subsetting the dataset by subjects in producing Figure 6.2, the raindrop scatter plots with 10 subjects but 20 eye records.

```
proc gplot data = iopcct;
  plot iop * cct = iop_type /HAXIS = AXIS2 VAXIS = AXIS1
noframe anno = annoa des = "Figure 6.1 Thunderstorm
Scatter Plot: All Data";
  FILENAME GSASFILE "&OUTLOC./Figure 6.1.&EXT";
  title1 "Figure 6.1 GAT and DCT IOP by CCT Value";
  title2 "Thunderstorm Scatter Plot: All Data from
100 Patients (200 Eyes)";
  footnote1 "&pgmpth.";
  label iop_type = 'Legend:';
run;

  plot iop * cct = iop_type /HAXIS = AXIS2 VAXIS = AXIS1
noframe anno = annoa (where = (1 < = recid < = 20)) des =
"Figure 6.2 Rain-drop Scatter Plot: Data from 10 Patients";
  FILENAME GSASFILE "&OUTLOC./Figure 6.2.&EXT";
  title1 "Figure 6.2 GAT and DCT IOP by CCT Value";
  title2 "Rain-drop Scatter Plot: Data from 10 Patients
(20 Eyes)";
  footnote1 &FMTFOOT "&pgmpth.";
  where 1001 < = subjid < = 1010;
```

```
     label iop_type = 'Legend:';
  run;
  quit;
```

6.3.2.1.3 *Producing Figure 6.1 Using the SGPLOT Procedure*

- Two Scatter statements are used to produce scatters for GAT and DCT IOPs by CCT.
- The *HIGHLOW* statement is used to connect the two IOPs for each eye with a straight line to form a raindrop. When the two IOPs for all eyes are connected with a "raindrop," a thunderstorm scatter plot is produced.

```
title1 "Figure 6.1. GAT and DCT IOP by CCT Value";
title2 "Thunderstorm Scatter Plot: All Data from 100
Patients (200 Eyes)";
footnote1 "&pgmpth.";
proc sgplot data = iopcct_hl noautolegend;
   scatter x = cct y = iop_gat/markerattrs = (symbol =
circle color = red size = 10) name = "GAT" legendlabel =
"GAT";
   scatter x = cct y = iop_dct/markerattrs = (symbol =
circlefilled color = black size = 10) name = "DCT"
legendlabel = "DCT";
   highlow x = cct high = iop_high low = iop_low/lineattrs
= (thickness = 2 PATTERN = solid COLOR = green);

   xaxis VALUES = (500 to 600 by 10) label = "CCT (microns)";
   yaxis VALUES = (10 to 40 by 5) label = "IOP (mm Hg)";
   keylegend "GAT" "DCT"/noborder title = 'Legend:';
run;
quit;

ods graphics/OUTPUTFMT = &outformat. imagename = "SGFig 6_2";
ods proclabel = "Rain-drop Scatter Plot: Data from 10
Patients (20 Eyes)";
title1 "Figure 6.2. GAT and DCT IOP by CCT Value";
title2 "Rain-drop Scatter Plot: Data from 10 Patients
(20 Eyes)";
footnote1 &FMTFOOT "&pgmpth.";
proc sgplot data = iopcct_hl noautolegend;
   where 1001 < = subjid < = 1010;
   scatter x = cct y = iop_gat/markerattrs = (symbol =
circle color = red size = 10) name = "GAT" legendlabel =
"GAT";
   scatter x = cct y = iop_dct / markerattrs = (symbol =
circlefilled color = black size = 10) name = "DCT"
legendlabel = "DCT";
```

```
  highlow x = cct high = iop_high low = iop_low/lineattrs
= (thickness = 2 PATTERN = solid COLOR = green);

  xaxis VALUES = (500 to 600 by 10) label = "CCT (microns)";
  yaxis VALUES = (10 to 40 by 5) label = "IOP (mm Hg)";
  keylegend "GAT" "DCT"/noborder title = 'Legend:';
run;
quit;
```

6.3.2.2 Main Sections and Features of the Second Program

6.3.2.2.1 Dataset Simulation

- A dataset is initially simulated to include 10 sites with 15 subjects at each site (*Site_Subj*), then 50 subjects are randomly dropped to allow a different number of subjects at each site to produce a dataset (*Site_Subj2*) with 100 subjects. The subject number at each site is counted using Proc Freq and numbers are saved in macro variables (n_site1, n_site2, etc.) using the call symput() function.

- Based on dataset *Site_subj2*, the IOP dataset is simulated to include the IOP values at hours 0, 2, and 8 in four visits—the baseline and postbaseline visits of weeks 4, 8, and 12. The IOP values are assigned based on a normal distribution with the preset mean and SD at each hour and visit.

6.3.2.2.2 Data Manipulation

- The IOP dataset is transposed for the ease of calculating the mean diurnal IOP at each visit, which is defined as the mean IOP value at hours 0, 2, and 8 of a visit.

- In order to display each subject on the x-axis, each subject is assigned an index number in the variable subj-index based on _N_, a system variable that contains the observation number of each subject in a dataset (*Site_Subj2*). Subjects in different sites are assigned an index with 1 unit higher than their previous site number to allow space to draw the reference lines in the figure. This is done in the following codes.

```
data subjid;
  set site_subj2;
  subj_id = _N_;
  if siteid = 1 then subj_index = _N_;
  if siteid = 2 then subj_index = _N_ + 1; ** leave
space for reference line;
  if siteid = 3 then subj_index = _N_ + 2;
  if siteid = 4 then subj_index = _N_ + 3;
  if siteid = 5 then subj_index = _N_ + 4;
```

```
      if siteid = 6 then subj_index = _N_ + 5;
      if siteid = 7 then subj_index = _N_ + 6;
      if siteid = 8 then subj_index = _N_ + 7;
      if siteid = 9 then subj_index = _N_ + 8;
      if siteid = 10 then subj_index = _N_ + 9;
      keep siteid subjid subj_index;
   run;
```

- Based on the subject numbers at each site (macro variables &n_site1, &n_site2, etc.), the reference line positions on the x-axis are saved in the macro variables 'Ref1', 'Ref2' etc. using *call symputx()*. The reference lines are used to separate each site on the x-axis.

- Datasets are merged to include mean diurnal IOP and subject index in one dataset (mndiur_all), and transposed to allow separate variables for mean diurnal IOP at each visit (mndiur_vst).

- The *Proc Format* procedure is used to include the format for the subject index. This procedure is used to suppress the display of subject index numbers on the x-axis by assigning all to none.

- A SAS annotated dataset (*anno_w12*) is created to draw a straight line between the mean diurnal IOP at the baseline and week 12 to form a raindrop for each subject.

6.3.2.2.3 *Producing Figure 6.3 Using Both GPLOT and SGPLOT Procedures*

As in the first SAS program to produce Figures 6.1 and 6.2, both GPLOT and SGPLOT procedures are used to produce Figure 6.3. The syntax and codes used are very similar to those discussed in the first SAS program.

6.3.2.3 **Main Sections and Features of the Third Program**

6.3.2.3.1 *Dataset Preparation*

A dataset is produced using the data from Liu's dissertation (Liu, 1998). In the dataset, numbers from 1 to 9 are used to represent the plant mineral nutrients of N, P, K, Ca, Mg, S, Cl, and Na. The numbers are used to position the nutrients on the x-axis. A format (*elemdf*) is created to display the nutrient names and salinity levels in the dataset and the plot.

6.3.2.3.2 *Producing Figure 6.4 Using Both GPLOT and SGPLOT Procedures*

Both GPLOT and SGPLOT procedures are used to produce Figure 6.4. The syntax and codes used are very similar to those discussed in the first SAS program.

6.4 Summary and Discussion

A thunderstorm/raindrop scatter plot is a useful data visualization tool to explore the overall data pattern and detect outliers for data with two or more values on the y-axis corresponding to one value on the x-axis. The values on the y-axis are usually continuous and those on the x-axis can be either continuous or discrete values. The thunderstorm scatter plots can be used in many areas for data with *before* and *after* measurements of the same end-point, including clinical research, agriculture, and finance.

The SAS annotate facility in GPLOT and the HIGHLOW statement in SGPLOT are used to produce raindrops by connecting two values on the y-axis with the same corresponding value on the x-axis. The SGPLOT might be easier to use to produce the thunderstorm/raindrop scatter plots because raindrops can be formed directly using the HIGHLOW statement without using the SAS annotate facility.

6.5 References

Allison, R. 2012. *SAS/GRAPH®: Beyond the Basics*. Cary, NC: SAS Institute, Inc.

Boehm, A.G., Weber, A., Pillunat, L.E., et al. 2008. "Dynamic Contour Tonometry in Comparison to Intracameral IOP Measurements," *Invest Ophthalmol Vis Sci.* 49: 2472–2477.

Cleveland, William. 1985. *The Elements of Graphing Data*. Cheshire, CT: Graphics Press.

Ehlers, N., Bramsen, T., and Sperling, S. 1975. "Applanation Tonometry and Central Corneal Thickness," *Acta Ophthalmol (Copenh)* 53: 34–43.

Goldmann, H., and Schmidt, T. 1957. "Uber Applanationstonometrie," *Ophthalmologica.* 134: 21–242.

Hyndman, R.J., and Shang, H.L. 2010. "Rainbow Plots, Bagplots, and Boxplots for Functional Data," *Journal of Computational and Graphical Statistics*. 19(1): 29–45.

Liu, C. 1998. "Effects of Humic Substances on Creeping Bentgrass Growth and Stress Tolerance." PhD diss., North Carolina State University.

Liu, C. 2013. "Introducing and Producing Thunderstorm or Rain-drop Scatter Plots Using the SAS/GRAPH® Annotate Facility." *SAS Global Forum 2013 Proceedings*. Cary, NC: SAS Institute Inc., http://support.sas.com/resources/papers/proceedings13/357-2013.pdf.

SAS Institute Inc. 2013. "SAS® 9.3 Product Documentation." SAS Knowledge Base, http://support.sas.com/documentation/93/index.html.

Tufte, Edward. 1983. *The Visual Display of Quantitative Information*. Cheshire, CT: Graphics Press.

Tufte, Edward. 1997. *Visual Explanations*. Cheshire, CT: Graphics Press.

Tufte, Edward. 2006. *Beautiful Evidence*. Cheshire, CT: Graphics Press.

Whitacre, M.M., and Stein, R. 1993. "Sources of Error with Use of Goldmann-Type Tonometers," *Surv Ophthalmol* 38: 1–30.

6.6 Appendix: SAS Programs for Producing the Sample Figures

The three SAS programs, Thunderstorm Scatter Plots_Example 1.sas, Thunderstorm Scatter Plots_Example 2.sas, and Thunderstorm Scatter Plots_Example 3.sas.sas, that were used to produce the four sample figures are presented in this section.

6.6.1 Thunderstorm Scatter Plots for Two Types of IOPs by Corresponding CCT

```
****************************************************************;
* Program Name: Thunderstorm Scatter Plots_Example 1.sas    *;
* Function: Produce the following two figures in both GPLOT *;
* and SGPLOT                                                *;
* -. Figure 6.1. GAT and DCT IOP by CCT Value:             *;
* - Thunderstorm Scatter Plot: All Data from 100 Patients  *;
* (200 Eyes)                                                *;
* -. Figure 6.2. GAT and DCT IOP by CCT Value:             *;
* - Rain-drop Scatter Plot: Data from 10 Patients (20 Eyes) *;
****************************************************************;
options mprint symbolgen nodate nonumber validvarname = v7
orientation = landscape;
%let pgmname = Thunderstorm Scatter Plots_Example 1.sas;
%let pgmloc = C:\SASBook\SAS Programs;
%let outloc = C:\SASBook\Sample Figures\Chapter 6;
%let pgmpth = &pgmloc.\&pgmname. &sysdate9. &systime. SAS
V&sysver.;
%let seed = 1699;
proc format;
   value eyedf
      1 = 'OD'
      2 = 'OS';
   value iopdf
      1 = 'GAT'
      2 = 'DCT';
run;
** Set the mean IOP for GAT and DCT for simulation purposes;
%let MN_GAT = 25;
%let MN_DCT = 26.5;

** Simulate 100 subject's IOP data in both eyes;
data iop;
   do i = 1 to 100; ** 100 subjects;
      do j = 1 to 2; ** 2 eyes;
         do k = 1 to 2; ** 2 IOP type;
            subjid = 1000 + i;
            eye = j;
```

```
      iop_type = k;
      if k = 1 then iop = round((RANNOR(&seed. + i + j + k)*
3.5 + &MN_GAT.),.1);
      else if k = 2 then iop = round((RANNOR(&seed. + i + j
+ k)* 3.5 + &MN_DCT.),.1);
      output;
      end;
    end;
  end;
  format eye eyedf. iop_type iopdf.;
  drop i j k;
run;

** Simulate 100 subjects' CCT data;
data cct;
  do i = 1 to 100; ** 100 subjects;
    do j = 1 to 2; ** 2 eyes;
      subjid = 1000 + i;
        eye = j;
        CCT = round((RANNOR(&seed. + i + j)* 50 + 550), 1);
        if 300 < = cct < 400 then cct = cct + 200;
** CCT between 500 to 600 microns;
        if 400 < = cct < 500 then cct = cct + 100;
        if 600 < cct < = 700 then cct = cct - 100;
        if 700 < cct < = 800 then cct = cct - 200;
        output;
    end;
  end;
  format eye eyedf.;
  drop i j;
run;

proc sort data = cct;
  by subjid eye;
run;

data iopcct;
  merge iop cct;
  by subjid eye;
run;

proc transpose data = iopcct out = iopcct2 prefix = iop_;
  by subjid eye cct;
  var iop;
  id iop_type;
run;

data iopcct2;
  set iopcct2;
  iop_diff = iop_dct - iop_gat;
```

```
     recid = _N_;
run;

** Annotate dataset to connect the 2 IOPs of the same eye by a
line;
data annoa;
  set iopcct2;
  by subjid;
     function = 'move'; xsys = '2'; ysys = '2'; x = cct; y =
iop_gat; output;
     function = 'draw'; xsys = '2'; ysys = '2'; color = 'VILG';
     x = cct; y = iop_dct; size = 2; line = 1; output;
run;

%LET FONTNAME = Times;
%LET DRIVER = PSCOLOR;%LEt EXT = PS;
goptions
   reset   = all
   GUNIT   = PCT
   rotate  = landscape
   gsfmode = replace
   gsfname = GSASFILE
   device  = &DRIVER
   lfactor = 1
   hsize   = 8 in
   horigin = 0 in
   vsize   = 6.5 in
   vorigin = 0 in
   ftext   = "&FONTNAME"
   htext   = 10pt
   ftitle  = "&FONTNAME"
   htitle  = 10pt
;

SYMBOL1 H = 2.5 C = RED CO = RED I = none W = 2 L = 1 font =
'albany amt/unicode' VALUE = '25cb'x;
SYMBOL2 H = 2.5 C = BLACK CO = BLACK I = none W = 2 L = 2 font
= 'albany amt/unicode' VALUE = '25cf'x;
AXIS1 MINOR = NONE OFFSET = (1,1) order = (10 to 40 by 5)
LABEL = (FONT = "&FONTNAME" ANGLE = 90 HEIGHT = 2.5 font =
"&FONTNAME" "IOP (mm Hg)") VALUE = (H = 1.5);
AXIS2 MINOR = NONE OFFSET = (1,1) order = (500 to 600 by 10)
LABEL = (FONT = "&FONTNAME" HEIGHT = 2.5 font = "&FONTNAME"
"CCT (microns)") VALUE = (H = 2);

** Upper-level boomkark description;
ods proclabel = "GAT and DCT IOP by CCT Value";
proc gplot data = iopcct;
```

```
  plot iop * cct = iop_type /HAXIS = AXIS2 VAXIS = AXIS1
noframe anno = annoa des = "Figure 6.1 Thunderstorm Scatter
Plot: All Data";
  FILENAME GSASFILE "&OUTLOC./Figure 6.1.&EXT";
  title1 "Figure 6.1 GAT and DCT IOP by CCT Value";
  title2 "Thunderstorm Scatter Plot: All Data from 100
Patients (200 Eyes)";
  footnote1 "&pgmpth.";
  label iop_type = 'Legend:';
  run;

  plot iop * cct = iop_type /HAXIS = AXIS2 VAXIS = AXIS1
noframe anno = annoa (where = (1 < = recid < = 20)) des =
"Figure 6.2 Rain-drop Scatter Plot: Data from 10 Patients";
  FILENAME GSASFILE "&OUTLOC./Figure 6.2.&EXT";
  title1 "Figure 6.2 GAT and DCT IOP by CCT Value";
  title2 "Rain-drop Scatter Plot: Data from 10 Patients
(20 Eyes)";
  footnote1 "&pgmpth.";
  where 1001 < = subjid < = 1010;
  label iop_type = 'Legend:';
  run;
quit;

**************************************************************;
** Reproducing the same figures using the sgplot procedure  *;
**************************************************************;
data iopcct_hl;
  set iopcct2;
    if iop_gat > = iop_dct then do;
    iop_high = iop_gat;
    iop_low = iop_dct;
  end;
  else do;
    iop_high = iop_dct;
    iop_low = iop_gat;
  end;
run;

%LET OUTPUTFMT = PS;
ods listing gpath = "&outloc.";
ods graphics/reset = all width = 8in height = 6in noborder
OUTPUTFMT = &OUTPUTFMT. imagename = "SGFig 6_1";
ods proclabel = "Thunderstorm Scatter Plot: All Data from
100 Patients (200 Eyes)";
title1 "Figure 6.1. GAT and DCT IOP by CCT Value";
title2 "Thunderstorm Scatter Plot: All Data from 100 Patients
(200 Eyes)";
footnote1 "&pgmpth.";
proc sgplot data = iopcct_hl noautolegend;
```

```
  scatter x = cct y = iop_gat/markerattrs = (symbol = circle
color = red size = 10) name = "GAT" legendlabel = "GAT";
  scatter x = cct y = iop_dct/markerattrs = (symbol =
circlefilled color = black size = 10) name = "DCT" legendlabel
= "DCT";
  highlow x = cct high = iop_high low = iop_low/lineattrs =
(thickness = 2 PATTERN = solid COLOR = green);
  xaxis VALUES = (500 to 600 by 10) label = "CCT (microns)";
  yaxis VALUES = (10 to 40 by 5) label = "IOP (mm Hg)";
  keylegend "GAT" "DCT"/noborder title = 'Legend:';
run;
quit;

ods graphics/OUTPUTFMT = &OUTPUTFMT. imagename = "SGFig 6_2";
ods proclabel = "Rain-drop Scatter Plot: Data from 10 Patients
(20 Eyes)";
title1 "Figure 6.2. GAT and DCT IOP by CCT Value";
title2 "Rain-drop Scatter Plot: Data from 10 Patients (20
Eyes)";
footnote1 "&pgmpth.";
proc sgplot data = iopcct_hl noautolegend;
  where 1001 < = subjid < = 1010;
  scatter x = cct y = iop_gat/markerattrs = (symbol = circle
color = red size = 10) name = "GAT" legendlabel = "GAT";
  scatter x = cct y = iop_dct/markerattrs = (symbol =
circlefilled color = black size = 10) name = "DCT" legendlabel
= "DCT";
  highlow x = cct high = iop_high low = iop_low/lineattrs =
(thickness = 2 PATTERN = solid COLOR = green);
  xaxis VALUES = (500 to 600 by 10) label = "CCT (microns)";
  yaxis VALUES = (10 to 40 by 5) label = "IOP (mm Hg)";
  keylegend "GAT" "DCT"/noborder title = 'Legend:';
run;
quit;
```

6.6.2 Thunderstorm Scatter Plots for IOP at Baseline and Week 12 by Subject

```
**************************************************************;
* Program Name: Thunderstorm Scatter Plots_Example 2.sas   *;
* Function: Produce the following two figures in both GPLOT *;
* and SGPLOT                                                *;
* - Figure 6.3. Thunderstorm Scatter Plot for Study Eye    *;
* Mean Diurnal IOP at Baseline and Week 12                 *;
**************************************************************;
options mprint symbolgen nodate nonumber validvarname = v7
orientation = landscape;
%let pgmname = Thunderstorm Scatter Plots_Example 2.sas;
```

```
%let pgmloc = C:\SASBook\SAS Programs\&pgmname. &sysdate9.
&systime. SAS V&sysver.;
%let outloc = C:\SASBook\Sample Figures\Chapter 6;

** set site number;
%let sitenum = 10;
%let seed = 1699;
%let SD = 3.5;

proc format;
  value visdf
    1 = 'Baseline'
    2 = 'Week 4'
    3 = 'Week 8'
    4 = 'Week 12';
  value hrdf
    1 = 'Hour 0'
    2 = 'Hour 2'
    3 = 'Hour 8';
  value sitedf
    0 - 100 = ' '
    OTHER = ' ';
run;

** Generate the required number of subjects;
data site_subj;
  do i = 1 to &sitenum.;
    do j = 1 to 15;
    siteid = i;
    subjid = i * 1000 + j; ** Subject id;
    shuffle = ranuni (&seed. + i + j);
    output;
    end;
  end;
  drop i j;
run;

proc sort data = site_subj;
  by shuffle;
run;

** Drop 50 subjects so each site might have different number
of subjects;
data site_subj2;
  set site_subj;
  if _N_ < = 50 then delete;
run;

proc sort data = site_subj2;
by siteid subjid;
```

```
run;

** Distribution of subjects by site;
proc freq data = site_subj2 noprint;
  table siteid/out = subj_freq (drop = PERCENT);
run;

** Save the subject number at each site to macro variables to
be used later-on;
data _NULL_;
  set subj_freq;
  call symputx('n_site'||left(put(siteid,best.)),
put(count, 2.0));
run;

** Set up the study eye IOP values per subject based on vis-
its/timepoints: assuming 6 to 8 mmHg reduction in IOP at
post-bsl visits;
data iop;
  set site_subj2;
  do i = 1 to 4; ** 4 visits;
    do j = 1 to 3; ** 3 timepoints/visit;
      visit = i;
      hour = j;
      if i = 1 then do; ** Baseline;
      if j = 1 then iop = round((RANNOR(&seed. + i + j)* &SD.
+ 25),.1);
        if j = 2 then iop = round((RANNOR(&seed. + i + j)* &SD.
+ 23),.1);
        if j = 3 then iop = round((RANNOR(&seed. + i + j)* &SD.
+ 22),.1);
    end;
    else if i > 1 then do;
        if j = 1 then iop = round((RANNOR(&seed. + i + j)* &SD.
+ 17.5),.1);
        if j = 2 then iop = round((RANNOR(&seed. + i + j)* &SD.
+ 16.5),.1);
        if j = 3 then iop = round((RANNOR(&seed. + i + j)* &SD.
+ 16.2),.1);
    end;
    output;
    end;
  end;
  drop i j shuffle;
  format visit visdf. hour hrdf.;
run;

proc sort data = iop;
  by siteid subjid visit hour;
run;
```

```
** Mean diurnal IOP at each visit: mean of hours 0, 2 and 8
average eye IOP at baseline;
proc transpose data = iop out = iop_t;
  by siteid subjid visit;
  var iop;
  id hour;
run;

data iop_MnDiur;
  set iop_t;
  iop_MnDiur = round (mean (Hour_0, Hour_2, Hour_8),.01);
  drop _NAME_;
run;

** Set the subject index number to be used in X-axis;
data subjid;
  set site_subj2;
  subj_id = _N_;
  if siteid = 1 then subj_index = _N_;
  if siteid = 2 then subj_index = _N_ + 1; ** leave space for
reference line;
  if siteid = 3 then subj_index = _N_ + 2;
  if siteid = 4 then subj_index = _N_ + 3;
  if siteid = 5 then subj_index = _N_ + 4;
  if siteid = 6 then subj_index = _N_ + 5;
  if siteid = 7 then subj_index = _N_ + 6;
  if siteid = 8 then subj_index = _N_ + 7;
  if siteid = 9 then subj_index = _N_ + 8;
  if siteid = 10 then subj_index = _N_ + 9;
  keep siteid subjid subj_index;
run;

** Set macro variables to hold the reference line locations on
the x-axis;
data _NULL_;
  refn1 = &n_site1. + 1;
  refn2 = refn1 + &n_site2. + 1;
  refn3 = refn2 + &n_site3. + 1;
  refn4 = refn3 + &n_site4. + 1;
  refn5 = refn4 + &n_site5. + 1;
  refn6 = refn5 + &n_site6. + 1;
  refn7 = refn6 + &n_site7. + 1;
  refn8 = refn7 + &n_site8. + 1;
  refn9 = refn8 + &n_site9. + 1;
  refn10 = refn9 + &n_site10. + 1;

  call symputx("Ref1", refn1);
  call symputx("Ref2", refn2);
  call symputx("Ref3", refn3);
  call symputx("Ref4", refn4);
```

```
    call symputx("Ref5", refn5);
    call symputx("Ref6", refn6);
    call symputx("Ref7", refn7);
    call symputx("Ref8", refn8);
    call symputx("Ref9", refn9);
    call symputx("Ref10", refn10);
run;

data mndiur_all;
  merge iop_MnDiur subjid;
  by siteid subjid;
run;

proc transpose data = MnDiur_all out = MnDiur_T prefix = vst_;
  by siteid subjid subj_index;
  var iop_MnDiur;
run;

** Mean Diurnal IOP at each visit;
data mndiur_vst;
  set MnDiur_T;
  drop _NAME_;
  rename vst_1 = BSL vst_2 = WK4 vst_3 = WK8 vst_4 = WK12;
run;

proc format;
  value indexid
    1 = ' '
     OTHER = ' ';
run;

** Annotate dataset to connect the 2 IOPs at baseline and;
** week 12 of the same subject by a straight line;
** (forming a rain drop);
data anno_w12;
  set mndiur_vst;
  by subj_index;
    function = 'move'; xsys = '2'; ysys = '2'; x = subj_index;
    y = bsl; output;
    function = 'draw'; xsys = '2'; ysys = '2'; color = 'VILG';
    x = subj_index; y = wk12;
    size = 2; line = 1; output;
    drop wk4 wk8;
run;

%LET FONTNAME = Times;
%LET DRIVER = PSCOLOR;%LEt EXT = PS;
goptions
  reset    = all
  GUNIT    = PCT
```

```
     rotate  = landscape
     gsfmode = replace
     gsfname = GSASFILE
     device  = &DRIVER
     hsize   = 8 in
     horigin = 0 in
     vsize   = 6.5 in
     vorigin = 0 in
     ftext   = "&FONTNAME"
     htext   = 10pt
     ftitle  = "&FONTNAME"
     htitle  = 10pt
;

SYMBOL1 H = 2.5 C = RED CO = RED I = none W = 2 L = 1 font =
'albany amt/unicode' VALUE = '25cb'x;
SYMBOL2 H = 2.5 C = BLACK CO = BLACK I = none W = 2 L = 2 font
= 'albany amt/unicode' VALUE = '25cf'x;
AXIS1 MINOR = NONE OFFSET = (1,1) order = (10 to 30 by 5)
LABEL = (FONT = "&FONTNAME" ANGLE = 90 HEIGHT = 2.5 font =
"&FONTNAME" "Mean Diurnal IOP at Baseline and Post-baseline
(mm Hg)") VALUE = (H = 1.5);
AXIS2 MAJOR = NONE MINOR = NONE OFFSET = (1,1) order = (0 to
110 by 1) LABEL = (FONT = "&FONTNAME" HEIGHT = 2.5 font =
"&FONTNAME" "Subject Grouped by Investigator Sites") VALUE =
(H = 2) reflabel = (position = bottom c = blue font =
"&FONTNAME" h = 2 j = r "1(n = &n_site1.))" "2(n = &n_site2.)"
"3(n = &n_site3.)" "4(n = &n_site4.)" "5(n = &n_site5.)"
"6(n = &n_site6.)" "7(n = &n_site7.)" "8(n = &n_site8.)"
"9(n = &n_site9.)" "10(n = &n_site10.)");
proc gplot data = mndiur_all;
  plot iop_MnDiur * subj_index = visit/HAXIS = AXIS2 VAXIS =
AXIS1 noframe anno = anno_w12 des = "Baseline and Week 12"
href = 0 &ref1. &ref2. &ref3. &ref4. &ref5. &ref6. &ref7.
&ref8. &ref9. &ref10.;
  FILENAME GSASFILE "&OUTLOC./Figure 6.3.&EXT.";
  title1 "Figure 6.3. Thunderstorm Scatter Plot for Mean
Diurnal IOP";
  title2 "Baseline and Week 12";
  footnote1 "&pgmpth.";
  where visit in (1, 4);
  label visit = 'Legend:'; format subj_index indexid.;
  run;
quit;

*************************************************************;
** Re-produce the figures using SGPLOT                   *;
*************************************************************;
proc transpose data = mndiur_all out = mndiur_all2;
  by subjid subj_index;
```

```
   var iop_MnDiur;
   id visit;
run;

%LET OUTPUTFMT = PS;
ods listing gpath = "&outloc.";
ods graphics/reset = all width = 8in height = 6in noborder
OUTPUTFMT = &OUTPUTFMT. imagename = "SGFig 6_3";
title1 "Figure 6.3. Thunderstorm Scatter Plot for Mean Diurnal
IOP";
title2 "Baseline and Week 12";
footnote1 "&pgmpth.";
proc sgplot data = mndiur_all2 noautolegend;
   scatter x = subj_index y = baseline/markerattrs = (symbol =
circle color = red size = 10) name = "BSL" legendlabel =
"Baseline";
   scatter x = subj_index y = week_12/markerattrs = (symbol =
circlefilled color = black size = 10) name = "PBSL"
legendlabel = "Week 12";
   highlow x = subj_index high = week_12 low = baseline/
lineattrs = (thickness = 2 PATTERN = solid COLOR = green);
   xaxis VALUES = (0 to 110 by 1) label = "Subject Grouped by
Investigator Sites" display = (noticks);
   yaxis VALUES = (10 to 30 by 5) label = "Mean Diurnal IOP at
Baseline and Week 12 (mm Hg)";
   REFLINE 0 &ref1. &ref2. &ref3. &ref4. &ref5. &ref6. &ref7.
&ref8. &ref9. &ref10./AXIS = x LABEL = ("1 (n = &n_site1.)"
"2 (n = &n_site2.)" "3 (n = &n_site3.)" "4(n = &n_site4.)"
"5 (n = &n_site5.)" "6 (n = &n_site6.)" "7 (n = &n_site7.)"
"8 (n = &n_site8.)" "9 (n = &n_site9.)" "10 (n = &n_site10.)")
labelpos = min labelloc = inside;
   format subj_index sitedf.;
   keylegend "BSL" "PBSL"/noborder title = 'Legend:';
run;
quit;
```

6.6.3 Raindrop Scatter Plots for Plant Nutrient Contents at Three Salinity Levels

```
*************************************************************;
* Program Name: Thunderstorm Scatter Plots_Example 3.sas   *;
* Function: Produce the following figure in both GPLOT and  *;
* SGPLOT                                                     *;
* - Figure 6.4. Rain-drop Scatter Plot for Plant Nutrient   *;
* Contents at Three Salinity Levels                         *;
*************************************************************;
options mprint symbolgen nodate nonumber validvarname = v7
orientation = landscape;
%let pgmname = Thunderstorm Scatter Plots_Example 3.sas;
```

```
%let pgmloc = C:\SASBook\SAS Programs\&pgmname. &sysdate9.
&systime. SAS V&sysver.;
%let outloc = C:\SASBook\Sample Figures\Chapter 6;

proc format;
  value elemdf
     1 = 'N'
     2 = 'P'
     3 = 'K'
     4 = 'Ca'
     5 = 'Mg'
     6 = 'S'
     7 = 'Cl'
     8 = 'Na'
     OTHER = ' ';
  value saltdf
     1 = 'None (0)'
     2 = 'Medium (8)'
     3 = 'High (16)';
run;

data nutrients;
  input salt element content;
  format element elemdf.;
  datalines;
  1 1 60.4
  2 1 58.2
  3 1 56.7
  1 2 6.24
  2 2 5.69
  3 2 5.04
  1 3 37.9
  2 3 26.5
  3 3 24.3
  1 4 6.25
  2 4 3.40
  3 4 3.41
  1 5 3.00
  2 5 3.38
  3 5 3.61
  1 6 5.35
  2 6 5.01
  3 6 4.99
  1 7 7.26
  2 7 16.9
  3 7 20.9
  1 8 0.69
  2 8 8.39
  3 8 9.23
  ;
run;
```

```
proc transpose data = nutrients out = nutrients_t prefix =
salt_;
   by element;
   var content;
   id salt;
run;

** Annotate dataset to connect the 2 IOPs of the same eye by
a line;
data anno_exp3;
   set nutrients_t;
     function = 'move'; xsys = '2'; ysys = '2'; x = element;
y = salt_1; output;
     function = 'draw'; xsys = '2'; ysys = '2'; color = 'VILG';
x = element;
     y = salt_2; size = 2; line = 1; output;
     function = 'draw'; xsys = '2'; ysys = '2'; color = 'CYAN';
x = element;
     y = salt_3; size = 2; line = 1; output;
run;

%LET FONTNAME = Times;
%LET DRIVER = PSCOLOR;%LEt EXT = PS;
goptions
   reset    = all
   GUNIT    = PCT
   rotate   = landscape
   gsfmode  = replace
   gsfname  = GSASFILE
   device   = &DRIVER
   lfactor  = 1
   hsize    = 8 in
   horigin  = 0 in
   vsize    = 6.5 in
   vorigin  = 0 in
   ftext    = "&FONTNAME"
   htext    = 10pt
   ftitle   = "&FONTNAME"
   htitle   = 10pt
;

SYMBOL1 H = 2.5 C = BLACK CO = BLACK I = none W = 2 L = 1 font
= 'albany amt/unicode' VALUE = '25cb'x;
SYMBOL2 H = 2.5 C = BLUE CO = BLUE      I = none W = 2 L = 2
font = 'albany amt/unicode' VALUE = '25b2'x;
SYMBOL3 H = 2.5 C = RED CO = RED I = none W = 2 L = 2 font =
'albany amt/unicode' VALUE = '25cf'x;
AXIS1 MINOR = NONE OFFSET = (1,1) order = (0 to 65 by 5) LABEL
= (FONT = "&FONTNAME" ANGLE = 90 HEIGHT = 2.5 font =
"&FONTNAME" "Nutrient Content (g/kg)") VALUE = (H = 1.5);
```

```
AXIS2 MAJOR = NONE MINOR = NONE OFFSET = (1,1) order = (0.5 to
8.5 by 0.5) LABEL = (FONT = "&FONTNAME" HEIGHT = 2.5 font =
"&FONTNAME" "Nutrient Name") VALUE = (H = 2);
FILENAME GSASFILE "&OUTLOC./Figure 6.4.&EXT";
title1 "Figure 6.4. Rain-drop Scatter Plot for Plant Nutrient
Contents at Three Salinity Levels";
footnote1 "&pgmloc.";


** Upper-level boomkark description;
ods proclabel = "Plant Nutrient by Salinity";
proc gplot data = nutrients;
  plot content * element = salt/HAXIS = AXIS2 VAXIS = AXIS1
noframe anno = anno_exp3 des = "Plant Nutrient by Salinity
Level";
  label salt = 'Salinity Level (ds/m EC):';
  format salt saltdf. element elemdf.;
    run;
quit;


%LET OUTPUTFMT = PS;
ods listing gpath = "&outloc.";
ods graphics/reset = all width = 8in height = 6in noborder
OUTPUTFMT = &OUTPUTFMT. imagename = "SGFig 6_4";
ods proclabel = "Plant Nutrient by Salinity";
title1 "Figure 6.4. Rain-drop Scatter Plot for Plant Nutrient
Contents at Three Salinity Levels";
footnote1 "&pgmloc.";
proc sgplot data = nutrients_t noautolegend;
  scatter x = element y = salt_1 / markerattrs = (symbol =
circle color = black size = 10) name = "None" legendlabel =
"None (0)";
  scatter x = element y = salt_2 / markerattrs = (symbol =
triangle color = blue size = 10) name = "Med" legendlabel =
"Medium (8)";
  scatter x = element y = salt_3 / markerattrs = (symbol =
circlefilled color = red size = 10) name = "High" legendlabel =
"High (16)";
  highlow x = element high = salt_2 low = salt_1/lineattrs =
(thickness = 2 PATTERN = solid COLOR = green);
  highlow x = element high = salt_3 low = salt_2/lineattrs =
(thickness = 2 PATTERN = solid COLOR = cyan);
  xaxis VALUES = (0.5 to 8.5 by 0.5) label = "Nutrient Name"
display = (noticks);
  yaxis VALUES = (0 to 65 by 5) label = "Nutrient Content (g/kg)";
  format element elemdf.;
  keylegend "None" "Med" "High"/noborder title = 'Salinity Level
(ds/m EC):';
run;
quit;
```

7

Spaghetti Plots

7.1 Introduction

In Chapter 3, we discussed how to produce line plots to display the mean value of the two treatment groups by time. Another interesting data visualization would be to show the actual IOP values for all subjects by time in the same figure, which is known as a spaghetti plot. A spaghetti plot displays many line plots in the same figure. When many lines are drawn together in the same figure, the lines appear like noodles or spaghetti, hence the name. Spaghetti plots allow readers to see the data pattern over time for all subjects in a sample, and are good visualization tools to examine the overall response of all subjects over time.

Spaghetti plots are especially useful to display data from a sample with a relatively small number of subjects, for example, in exploratory phase I or II clinical studies.

7.2 Application Examples

To illustrate the application and production of spaghetti plots, three sample figures are presented in this chapter based on clinical research in the glaucoma therapeutic area. Let's design a virtual clinical trial to compare the effects of a New Drug to those of an Active Control on intraocular pressure (IOP) reduction in patients with glaucoma or ocular hypertension (OHT). The IOP values are measured at three time-points of 8 a.m., 10 a.m., and 4 p.m. each day on the baseline visit, and on days 1, 3, and 7. There are 50 patients enrolled and randomized to either the New Drug or the Active Control group at the 1:1 ratio. Subjects take the study medications after the baseline visit.

Sample Figures 7.1 and 7.2 are spaghetti plots showing the actual IOP values at 8 a.m., 10 a.m., and 4 p.m. of each visit for subjects in the Active Control and New Drug group, respectively. Figure 7.3 positions the two spaghetti plots for the two treatment groups (Figures 7.1 and 7.2) in one page, which allows

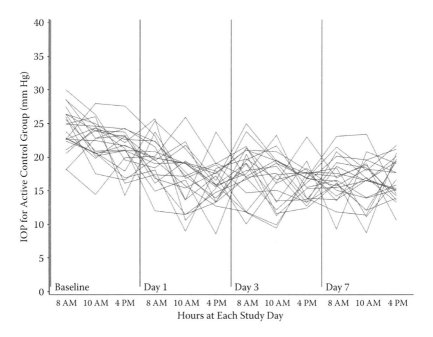

FIGURE 7.1
Spaghetti plot of IOP for individual subject in the Active Control group ($N = 22$).

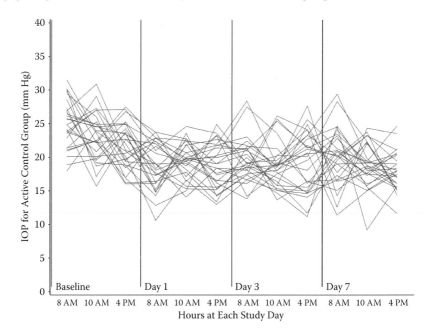

FIGURE 7.2
Spaghetti plot of IOP for individual subject in the New Drug group ($N = 28$).

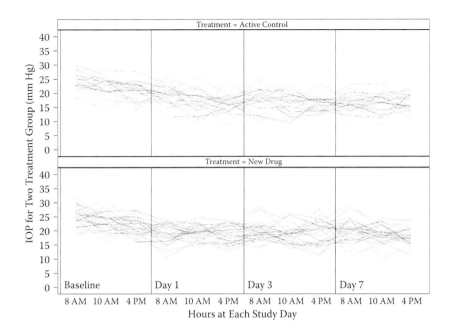

FIGURE 7.3
Spaghetti plot of IOP for individual subject (produced using SGPANEL), with the two treatments displayed on one page.

easier visualization and comparison of the IOP values for the two treatment groups. Based on the simulated data, subjects in the New Drug group have lower IOP values than those in the Active Control group overall, indicating that the New Drug reduced IOP values better than the Active Control.

The basic concepts and techniques described in this chapter for producing spaghetti plots can be applied to other clinical and nonclinical research areas as well.

7.3 Producing the Sample Figures

7.3.1 Data Structure and SAS Annotated Dataset

Table 7.1 displays the structure of the IOP data for two subjects. Each subject has 12 records with IOP values at 3 time points on each of the 4 visits. There are 50 subjects with a total of 600 records in the simulated dataset called IOP.

In PROC SGPLOT, a statistical graphic (SG) annotation dataset (anno_label) is used to position the labels for visits (Baseline, Day 1, etc.) in the plot. In the annotation dataset (Table 7.2), both the X1SPACE and Y1SPACE use the actual

TABLE 7.1

IOP Data for Two Subjects

Treatment	Subjid	Visit	Hour	IOP	Time
Active Control	1001	Baseline	8 a.m.	26.4	1
Active Control	1001	Baseline	10 a.m.	23.9	2
Active Control	1001	Baseline	4 p.m.	23	3
Active Control	1001	Day 1	8 a.m.	17.1	4
Active Control	1001	Day 1	10 a.m.	11.5	5
Active Control	1001	Day 1	4 p.m.	14.8	6
Active Control	1001	Day 3	8 a.m.	16.7	7
Active Control	1001	Day 3	10 a.m.	19.1	8
Active Control	1001	Day 3	4 p.m.	16.9	9
Active Control	1001	Day 7	8 a.m.	15.7	10
Active Control	1001	Day 7	10 a.m.	18.2	11
Active Control	1001	Day 7	4 p.m.	17.7	12
New Drug	1002	Baseline	8 a.m.	29.5	1
New Drug	1002	Baseline	10 a.m.	26	2
New Drug	1002	Baseline	4 p.m.	16.9	3
New Drug	1002	Day 1	8 a.m.	12.8	4
New Drug	1002	Day 1	10 a.m.	15	5
New Drug	1002	Day 1	4 p.m.	24.9	6
New Drug	1002	Day 3	8 a.m.	19	7
New Drug	1002	Day 3	10 a.m.	20.2	8
New Drug	1002	Day 3	4 p.m.	24.1	9
New Drug	1002	Day 7	8 a.m.	12.6	10
New Drug	1002	Day 7	10 a.m.	24.4	11
New Drug	1002	Day 7	4 p.m.	23.5	12

TABLE 7.2

SG Annotate Dataset Used in SGPLOT Procedure to Display the Visit Labels

Function	x1	y1	x1space	y1space	anchor	label	textcolor
text	0.5	1	datavalue	datavalue	left	Baseline	green
text	3.5	1	datavalue	datavalue	left	Day 1	green
text	6.5	1	datavalue	datavalue	left	Day 3	green
text	9.5	1	datavalue	datavalue	left	Day 7	green

data values with X1 and Y1 as the coordinates. The value of the FUNCTION variable (TEXT) tells the program to draw the labels ("Baseline," "Day 1," etc.) in the specified X1 and Y1 coordinates.

In using PROC SGPANEL to produce Figure 7.3 with the two spaghetti plots displayed on the same page, the DATAVALUE in X1PSACE and Y1SPACE variables do not work, and GRAPHPERCENT is used instead in the SG annotation dataset (ano_label2 in Table 7.3). Please note that the X1

TABLE 7.3

SG Annotate Dataset Used in SGPANEL Procedure to Display the Visit Labels

Function	x1	y1	x1space	y1space	anchor	label	textcolor
text	10	15	GRAPHPERCENT	GRAPHPERCENT	left	Baseline	green
text	35	15	GRAPHPERCENT	GRAPHPERCENT	left	Day 1	green
text	55	15	GRAPHPERCENT	GRAPHPERCENT	left	Day 3	green
text	80	15	GRAPHPERCENT	GRAPHPERCENT	left	Day 7	green

and Y1 values in Table 7.3 are different than those in Table 7.2. The X1 and Y1 values represent the location as a percentage of each axis (e.g., 10% in x-axis and 5% in y-axis, etc.) for the labels to be positioned in the figure.

For a more detailed introduction to SG annotation, please refer to the *SAS 9.3 ODS Graphics: Procedures Guide* (SAS Institute Inc., 2012).

7.3.2 Notes to SAS Programs

The SAS programs used to produce the sample plots using GPLOT and GREPLAY in SAS/GRAPH, and SGPLOT and SGPANEL in ODS Graphics are provided in Section 7.6. The programs consist of the following main sections.

7.3.2.1 Dataset Simulation and Manipulation

- IOP data for 50 subjects are simulated from a normal distribution with the preset mean and standard deviation (SD). Subjects are randomly assigned to the Active Control or New Drug group at the 1:1 ratio.
- The subject numbers in each treatment group are calculated using PROC FREQ and saved in macro variables using the function CALL SYMPUT(). The macro variables are used to display the subject numbers for each treatment group.
- A new variable *TIME* is created to represent the 3 hours in 4 visits, to be used on the x-axis.

7.3.2.2 Producing the Sample Figures Using the GPLOT and GREPLAY Procedures

- PROC GPLOT is used to produce the spaghetti plots in Figures 7.1 and 7.2 for the two treatment groups. Spaghetti plots are produced by using *PLOT IOP*TIME = SUBJID* together with the *INTERPOL = JOIN* option in the SYMBOL statement.
- The JOIN option in INTERPOL joins all the data points together with a solid line (line type specified by using LINE = 1), and the same SYMBOL definition is used for all the subjects in the treatment

group (= &N_Trt1.). When line plots for all subjects in the treatment group are displayed in the same figure, a spaghetti plot is created.

- The GOUT option is used to specify the SAS category (Spaghetti) to save the graphics output (Ttrt1 and Trt2). If no libref is specified, SAS would create the category (Spaghetti) in the WORK temporary library.
- PROC GREPLAY is used to replay the two stored graphs (Trt1 and Trt2) in the same page using the SAS-provided template V2 from a template catalog located at SASHELP.TEMPLT.

```
SYMBOL H = 2 C = gray I = join LINE = 1 repeat = &N_Trt1.;
axis1 major = none minor = none order = (0.5 to 12.5 by.5)
   label = (h = 2.5 font = "&fontname" "Hours at Each
Study Day")
   reflabel = (position = bottom c = green font =
"&fontname" h = 2.5 j = r
   "Baseline" "Day 1" "Day 3" "Day 7");
axis2 minor = none order = (0 to 40 by 5) label = (a = 90
r = 0 h = 2.5
   font = "&fontname" "IOP For Active Control Group (mm
Hg)");
title1 "Figure 7.1 Spaghetti Plot of IOP for All
Subjects";
title2 "Active Control (N = &N_Trt1.)";
footnote1 "&pgmpth.";
FILENAME GSASFILE "&OUTLOC.\Figure 7.1.&EXT";
proc gplot data = iop_time gout = goutloc.Spaghetti;
   plot iop*time = subjid/vaxis = axis2 haxis = axis1
hminor = 1 vminor = 1 noframe
      caxis = STGB nolegend href = 0.5 href = 3.5 href = 6.5
href = 9.5 name = "Trt1";
   where trtnum = 1;
   format time timedf.;
run;

axis2 minor = none order = (0 to 40 by 5) label = (a = 90
r = 0 h = 2.5
   font = "&fontname" "IOP for New Drug Group (mm Hg)");
SYMBOL H = 2 C = gray I = join LINE = 1 repeat =
&N_Trt2.;

title1 "Figure 7.2 Spaghetti Plot of IOP for All
Subjects";
title2 "New Drug (N = &N_Trt2.)";
footnote1 "&pgmpth.";
FILENAME GSASFILE "&OUTLOC.\Figure 7.2.&EXT";
   plot iop*time = subjid/vaxis = axis2 haxis = axis1
hminor = 1 vminor = 1 noframe
      caxis = STGB nolegend href = 0.5 href = 3.5 href =
6.5 href = 9.5 name = "Trt2";
```

```
   where trtnum = 2;
   format time timedf.;
run;

title1 "Figure 7.3 Spaghetti Plot of IOP for All
Subjects";
title2 "With Two Treatments Displayed in One Page";
footnote1 "&pgmpth.";
FILENAME GSASFILE "&OUTLOC.\Figure 7.3.&EXT";
proc greplay igout = goutloc.Spaghetti;
   tc = sashelp.templt;
   template = v2;
   ** Replay into the chosen template;
   treplay 1:TRT1 2:TRT2;
run;
quit;
```

- If you want to see the list of templates stored in SASHELP.TEMPLT, you can run the following codes. The code list TC lists the templates in the category in the SAS LOG window and preview v2 creates a preview of the V2 template in the Results window.

```
proc greplay nofs tc = sashelp.templt;
   list tc;
   preview v2;
run; quit;
```

7.3.2.3 Producing the Sample Figures Using the SGPLOT and SGPANEL Procedures

- PROC SGPLOT is used to reproduce the spaghetti plots in Figures 7.1 and 7.2 for the two treatment groups. Spaghetti plots are produced by using the SERIES plot statement with the GROUP = SUBJID option. The SG annotation dataset "anno_label" is used to place the labels for visits in the specified positions.
- PROC SGPANEL is used to reproduce Figure 7.3 with spaghetti plots for the two treatment groups displayed in a two-row panel (up-down layout). The SG annotation dataset "anno_label2" is used to place the labels for visits in the specified positions.

```
title1 "Figure 7.1 Spaghetti Plot of IOP for All Subjects";
title2 "Active Control (N = &N_Trt1.)";
footnote1 "&pgmpth.";
```

```
proc sgplot data = iop_time noautolegend sganno = anno_label;
   where trtnum = 1;
   series x = time y = iop/group = subjid lineattrs =
(color = gray pattern = solid);
   refline 0.5 3.5 6.5 9.5/axis = x lineattrs = (color =
black pattern = solid);
   xaxis values = (0.5 to 12.5 by.5) label = "Hours at
Each Study Day"
      tickvalueformat = timedf. display = (noticks);
   yaxis values = (0 to 40 by 5) label = "IOP For Active
Control Group (mm Hg)";
run;
quit;

ods graphics/reset = all width = 8in height = 6in noborder
OUTPUTFMT = &OUTPUTFMT. imagename = "SGFig7_2";
title1 "Figure 7.2 Spaghetti Plot of IOP for All Subjects";
title2 "New Drug (N = &N_Trt2.)";
footnote1 "&pgmpth.";
proc sgplot data = iop_time noautolegend sganno =
anno_label;
   where trtnum = 2;
   series x = time y = iop/group = subjid lineattrs =
(color = gray pattern = solid);
   refline 0.5 3.5 6.5 9.5/axis = x lineattrs = (color =
black pattern = solid);
   xaxis values = (0.5 to 12.5 by.5) label = "Hours at
Each Study Day"
      tickvalueformat = timedf. display = (noticks);
   yaxis values = (0 to 40 by 5) label = "IOP For New Drug
Group (mm Hg)";
run;
quit;

** Put the two figues by treatment group in one page
using SGPANEL;
ods listing gpath = "&outloc.";
ods graphics/reset = all width = 8in height = 6in
noborder OUTPUTFMT = &OUTPUTFMT. imagename = "SGFig7_3";

title1 "Figure 7.3 Spaghetti Plot of IOP for All Subjects";
title2 "With Two Treatments Displayed in One Page";
footnote1 "&pgmpth.";
proc sgpanel data = iop_time noautolegend sganno =
anno_label2;
   panelby trtnum/rows = 2;
   series x = time y = iop/group = subjid lineattrs =
(color = gray pattern = solid);
   refline 0.5 3.5 6.5 9.5/axis = x lineattrs = (color =
black pattern = solid);
```

```
    COLAXIS values = (0.5 to 12.5 by.5) label = "Hours at
Each Study Day"
      tickvalueformat = timedf. display = (noticks);
    rowaxis values = (0 to 40 by 5) label = "IOP for Two
Treatment Groups (mm Hg)";
run;
quit;
```

7.4 Summary and Discussion

Both the GPLOT and SGPLOT procedures can produce spaghetti plots. The features, including the pros and cons of using the GPLOT and SGPLOT, are summarized in Table 7.4. In GPLOT, the spaghetti plot is produced

TABLE 7.4

Comparing PROC GPLOT and PROC SGPLOT in Producing Spaghetti Plots

Features	GPLOT and GREPLAY in SAS/GRAPH	SGPLOT and SGPANEL in ODS Graphics
Spaghetti Lines	Using a *"PLOT x*y = subjid"* statement together with *"INTERPOL = Join"* option in the SYMBOL statement	Using a SERIES plot statement with the GROUP option: *series X = time Y = response/GROUP = subjid;*
Line and Marker Attributes	Using SYMBOL statement	Using LINEATTRS and MARKERATTRS plot statements
Axis Attributes	AXIS global statements and VAXIS/HAXIS plot statement options	XAXIS and YAXIS plot statements
Legend	Global LEGEND statement	LEGENDLABEL in MARKERS plot statement
Reference Lines and Labels	HREF and VREF option in plot statement Note: Reference labels are put in the right side of the reference lines by using *"j = r (justify = right)"* option.	AXIS = option in REFLINE plot statement. Reference label is put in the middle position of a reference line automatically, and the justify option is not available. Annotated facility can be used to display the reference label in the preferred position and color.
Multipanel Figure	Using the GREPLAY procedure to replay the graphs saved in the SAS category to two panels. The panels have their own x- and y-axes.	Using the SGPANEL procedure to place the spaghetti plots in two panels. The two panels share the same x-axis in up-down layout with 2 rows.
Pros	Easier to handle and position the reference line labels	Easier to use SGPANEL to produce multi-panel plot.
Cons	Multipanel plots are produced using REPLAY with original figures saved in an output category.	Reference line labels are handled and positioned using the SG annotate facility.

using a *PLOT X*Y = subjid* statement together with the *INTERPOL = Join* option in the SYMBOL statement. In SGPLOT, the spaghetti plot can be produced using a SERIES statement with the GROUP option (*Series X = time Y = response/GROUP = subjid*). The visit labels can easily be positioned by specifying the REFLABEL in the AXIS statement in GPLOT. They are placed using the SG annotation dataset in SGPLOT.

The GRELAY procedure in SAS/GRAPH can be used to replay, reformat, and reuse the graphs that are stored in an SAS category. However, this can be more easily done using the SGPANEL procedure in ODS Graphics. The two graphs, placed in two panels in the up and down layout in SGPANEL, also offer better visualization because they share the same x-axis and the same labels for both axes.

7.5 References

SAS Institute Inc. 2012. *SG Annotation. SAS® 9.3 ODS Graphics: Procedures Guide,* 3rd ed. Cary, NC: SAS Institute Inc.

7.6 Appendix: SAS Programs for Producing the Sample Figures

```
***************************************************************;
* Program Name: Chapter 7 Spaghetti Plots.sas              *;
* Descriptions: Producing the following sample figures in  *;
* chapter 7                                                *;
* - Figure 7.1 Spaghetti Plot of IOP for All Subjects:     *;
* Active Control                                           *;
* - Figure 7.2 Spaghetti Plot of IOP for All Subjects:     *;
* New Drug                                                 *;
* - Figure 7.3 Spaghetti Plot of IOP for All Subjects:     *;
* With Two Treatments Displayed in One Page                *;
***************************************************************;
options mprint symbolgen nodate nonumber validvarname = v7
orientation = landscape;
%let pgmname = Chapter 7 Spaghetti Plots.sas;
%let outloc = C:\SASBook\Sample Figures\Chapter 7;
%let pgmloc = C:\SASBook\SAS Programs;
%let pgmpth = &pgmloc.\&pgmname. &sysdate9. &systime. SAS
V&sysver.;
libname goutloc "C:\SASBook\Sample Figures\Chapter 7";
** Set-up the site, subject number and SD for data simulation;
%let subjnum = 50;
```

```
%let sd = 3.5;
%let seed = 07;
proc format;
   value trtdf
      1 = 'Active Control'
      2 = 'New Drug'
      OTHER = ' ';
   value visdf
      1 = 'Baseline'
      2 = 'Day 1'
      3 = 'Day 3'
      4 = 'Day 7';
   value timedf
      1, 4, 7, 10 = '8 AM'
      2, 5, 8, 11 = '10 AM'
      3, 6, 9, 12 = '4 PM'
      OTHER = ' ';
run;

** Generate the required number of subjects and randomly
assign to two treatments;
data subj;
   label trtnum = "Treatment";
   do i = 1 to &subjnum.;
      subjid = 1000+ i;
      if ranuni (&seed.) < = 0.5 then trtnum = 1;
         else trtnum = 2;
      output;
   end;
   drop i;
run;

proc freq data = subj noprint;
   table trtnum/out = subj_trt;
run;

** Save the subject number at each treatment group to macro
variables for later use;
data _null_;
   set subj_trt;
   if trtnum = 1 then call symput ("N_Trt1", put(count, 3.0));
   if trtnum = 2 then call symput ("N_Trt2", put(count, 3.0));
run;

** Set up the IOP Values based on the trt assignment and
visits/timepoints;
data iop;
   set subj;
   do i = 1 to 4; ** 4 visits;
```

```
   do j = 1 to 3; ** 3 timepoints/visit;
      visit = i;
      hour = j;
      if i = 1 then do; ** Baseline;
         if j = 1 then iop = round((RANNOR(&seed.)* &SD.
+ 25),.1); ** Hour 0;
         if j = 2 then iop = round((RANNOR(&seed.)* &SD.
+ 23),.1); ** Hour 2;
         if j = 3 then iop = round((RANNOR(&seed.)* &SD.
+ 22),.1); ** Hour 8;
      end;
      else if i > 1 and trtnum = 1 then do; ** Post-baseline:
active trt;
         if j = 1 then iop = round((RANNOR(&seed.)* &SD.
+ 17.5),.1); ** Hour 0;
         if j = 2 then iop = round((RANNOR(&seed.)* &SD.
+ 16.5),.1); ** Hour 2;
         if j = 3 then iop = round((RANNOR(&seed.)* &SD.
+ 16.2),.1); ** Hour 8;
      end;
      else if i > 1 and trtnum = 2 then do; ** Post-baseline:
active control;
         if j = 1 then iop = round((RANNOR(&seed.)* &SD.
+ 20),.1); ** Hour 0;
         if j = 2 then iop = round((RANNOR(&seed.)* &SD.
+ 19),.1); ** Hour 2;
         if j = 3 then iop = round((RANNOR(&seed.)* &SD.
+ 18.7),.1); ** Hour 8;
      end;
      output;
   end;
  end;
  drop i j;
  format visit visdf. hour timedf. trtnum trtdf.;
run;
proc sort data = iop;
  by subjid visit hour;
run;
data iop_time;
  set iop;
  if visit = 1 and hour = 1 then time = 1;
  if visit = 1 and hour = 2 then time = 2;
  if visit = 1 and hour = 3 then time = 3;
  if visit = 2 and hour = 1 then time = 4;
  if visit = 2 and hour = 2 then time = 5;
  if visit = 2 and hour = 3 then time = 6;
  if visit = 3 and hour = 1 then time = 7;
  if visit = 3 and hour = 2 then time = 8;
  if visit = 3 and hour = 3 then time = 9;
  if visit = 4 and hour = 1 then time = 10;
```

```
   if visit = 4 and hour = 2 then time = 11;
   if visit = 4 and hour = 3 then time = 12;
run;
proc sort data = iop_time;
   by subjid time;
run;

******************************************************;
* Produce the sample figures using GPLOT and GREPLAY    *;
******************************************************;
%LET FONTNAME = Times;
%LET DRIVER = PSCOLOR;%LEt EXT = PS;
goptions
   reset = all
   GUNIT = PCT
   rotate = landscape
   gsfmode = replace
   gsfname = GSASFILE
   device = &DRIVER
   lfactor = 1
   hsize = 8 in
   horigin = 0 in
   vsize = 6 in
   vorigin = 0 in
   ftext = "&FONTNAME"
   htext = 10pt
   ftitle = "&FONTNAME"
   htitle = 10pt
;

SYMBOL H = 2 C = gray I = join LINE = 1 repeat = &N_Trt1.;
axis1 major = none minor = none order = (0.5 to 12.5 by.5)
   label = (h = 2.5 font = "&fontname" "Hours at Each Study Day")
   reflabel = (position = bottom c = green font = "&fontname"
h = 2.5 j = r
   "Baseline" "Day 1" "Day 3" "Day 7");
axis2 minor = none order = (0 to 40 by 5) label = (a = 90 r = 0
h = 2.5
   font = "&fontname" "IOP For Active Control Group (mm Hg)");
title1 "Figure 7.1 Spaghetti Plot of IOP for All Subjects";
title2 "Active Control (N = &N_Trt1.)";
footnote1 "&pgmpth.";
FILENAME GSASFILE "&OUTLOC.\Figure 7.1.&EXT";
proc gplot data = iop_time gout = goutloc.Spaghetti;
   plot iop*time = subjid/vaxis = axis2 haxis = axis1 hminor = 1
vminor = 1 noframe
      caxis = STGB nolegend href = 0.5 href = 3.5 href = 6.5
href = 9.5 name = "Trt1";
   where trtnum = 1;
   format time timedf.;
run;
```

```
axis2 minor = none order = (0 to 40 by 5) label = (a = 90 r = 0
h = 2.5
   font = "&fontname" "IOP for New Drug Group (mm Hg)");
SYMBOL H = 2 C = gray I = join LINE = 1 repeat = &N_Trt2.;
title1 "Figure 7.2 Spaghetti Plot of IOP for All Subjects";
title2 "New Drug (N = &N_Trt2.)";
footnote1 "&pgmpth.";
FILENAME GSASFILE "&OUTLOC.\Figure 7.2.&EXT";
   plot iop*time = subjid/vaxis = axis2 haxis = axis1 hminor = 1
vminor = 1 noframe
     caxis = STGB nolegend href = 0.5 href = 3.5 href = 6.5
href = 9.5 name = "Trt2";
   where trtnum = 2;
   format time timedf.;
run;

title1 "Figure 7.3 Spaghetti Plot of IOP for All Subjects";
title2 "With Two Treatments Displayed in One Page";
footnote1 "&pgmpth.";
FILENAME GSASFILE "&OUTLOC.\Figure 7.3.&EXT";
proc greplay igout = goutloc.Spaghetti;
   tc = sashelp.templt;
   template = v2;
   ** Replay into the chosen template;
   treplay 1:TRT1 2:TRT2;
run;
quit;

************************************************************;
** Reproduce the same figures using SGPLOT and SGPANEL    *;
************************************************************;
** SG annotate to label the visits of the reference lines;
data anno_label;
   function = "text"; x1 = 0.5; y1 = 1; x1space = "datavalue";
   y1space = "datavalue"; anchor = "left";
   label = "Baseline"; textcolor = "green"; output;

   function = "text"; x1 = 3.5; y1 = 1; x1space = "datavalue";
   y1space = "datavalue"; anchor = "left";
   label = "Day 1"; textcolor = "green"; output;

   function = "text"; x1 = 6.5; y1 = 1; x1space = "datavalue";
   y1space = "datavalue"; anchor = "left";
   label = "Day 3"; textcolor = "green"; output;

   function = "text"; x1 = 9.5; y1 = 1; x1space = "datavalue";
   y1space = "datavalue"; anchor = "left";
   label = "Day 7"; textcolor = "green"; output;
run;
```

```
data anno_label2;
   function = "text"; x1 = 10; y1 = 15; x1space = "GRAPHPERCENT";
   y1space = "GRAPHPERCENT"; anchor = "left";
   label = "Baseline"; textcolor = "green"; output;
   function = "text"; x1 = 35; y1 = 15; x1space = "GRAPHPERCENT";
   y1space = "GRAPHPERCENT"; anchor = "left";
   label = "Day 1"; textcolor = "green"; output;
   function = "text"; x1 = 55; y1 = 15; x1space = "GRAPHPERCENT";
   y1space = "GRAPHPERCENT"; anchor = "left";
   label = "Day 3"; textcolor = "green"; output;
   function = "text"; x1 = 80; y1 = 15; x1space = "GRAPHPERCENT";
   y1space = "GRAPHPERCENT"; anchor = "left";
   label = "Day 7"; textcolor = "green"; output;
run;

%LET OUTPUTFMT = PS;
ods listing gpath = "&outloc.";
ods graphics/reset = all width = 8in height = 6in noborder
OUTPUTFMT = &OUTPUTFMT. imagename = "SGFig7_1";

title1 "Figure 7.1 Spaghetti Plot of IOP for All Subjects";
title2 "Active Control (N = &N_Trt1.)";
footnote1 "&pgmpth.";
proc sgplot data = iop_time noautolegend sganno = anno_label;
   where trtnum = 1;
   series x = time y = iop/group = subjid lineattrs = (color =
gray pattern = solid);
   refline 0.5 3.5 6.5 9.5/axis = x lineattrs = (color = black
pattern = solid);
   xaxis values = (0.5 to 12.5 by.5) label = "Hours at Each
Study Day"
      tickvalueformat = timedf. display = (noticks);
   yaxis values = (0 to 40 by 5) label = "IOP For Active Control
Group (mm Hg)";
run;
quit;

ods graphics/reset = all width = 8in height = 6in noborder
OUTPUTFMT = &OUTPUTFMT. imagename = "SGFig7_2";
title1 "Figure 7.2 Spaghetti Plot of IOP for All Subjects";
title2 "New Drug (N = &N_Trt2.)";
footnote1 "&pgmpth.";
proc sgplot data = iop_time noautolegend sganno = anno_label;
   where trtnum = 2;
   series x = time y = iop/group = subjid lineattrs = (color =
gray pattern = solid);
   refline 0.5 3.5 6.5 9.5/axis = x lineattrs = (color = black
pattern = solid);
```

```
  xaxis values = (0.5 to 12.5 by.5) label = "Hours at Each
Study Day"
     tickvalueformat = timedf. display = (noticks);
  yaxis values = (0 to 40 by 5) label = "IOP For New Drug Group
(mm Hg)";
run;
quit;

** Put the two figues by treatment group in one page using
SGPANEL;
ods listing gpath = "&outloc.";
ods graphics/reset = all width = 8in height = 6in noborder
OUTPUTFMT = &OUTPUTFMT. imagename = "SGFig7_3";
title1 "Figure 7.3 Spaghetti Plot of IOP for All Subjects";
title2 "With Two Treatments Displayed in One Page";
footnote1 "&pgmpth.";
proc sgpanel data = iop_time noautolegend sganno = anno_label2;
  panelby trtnum/rows = 2;
  series x = time y = iop/group = subjid lineattrs = (color =
gray pattern = solid);
  refline 0.5 3.5 6.5 9.5/axis = x lineattrs = (color = black
pattern = solid);
  COLAXIS values = (0.5 to 12.5 by.5) label = "Hours at Each
Study Day"
     tickvalueformat = timedf. display = (noticks);
  rowaxis values = (0 to 40 by 5) label = "IOP for Two
Treatment Groups (mm Hg)";
run;
quit;
```

8

Bar Charts

8.1 Introduction

A bar chart is a graph with rectangular bars with lengths that are proportional to the values that they represent. The bars can be plotted vertically or horizontally (Wikipedia). A bar chart is an alternative to line plots to display the mean or midpoint response when the response is categorized by a discrete value. Bar charts are usually used to compare a response value among specific categories.

Bar charts have applications in many areas, including clinical research, agriculture, and education.

8.2 Application Examples

To illustrate the application and production of bar charts, two examples with three sample figures are presented in this chapter based on clinical research in the glaucoma therapeutic area.

8.2.1 Example 1: Bar Chart of Drug Effects on IOP Reduction by Baseline IOP Categories

It is commonly felt that a drug's effect on a patient's intraocular pressure (IOP) reduction is influenced by the patient's baseline or initial IOP values. Bar charts of a drug's effects on IOP reduction by different baseline IOP categories (< 24, 24 to 26, and > 26 mm Hg) are shown in Figures 8.1 and 8.2. Figure 8.1 allows readers to examine the drug's effect at week 12 for each baseline IOP category, and Figure 8.2 displays the effects at all three post-baseline visits. Figure 8.2 is a classification bar chart and is produced using GREPLAY in SAS/GRAPH and in SGPANEL using the PANELBY statement in ODS Graphics.

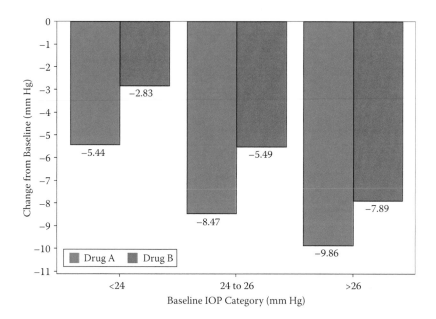

FIGURE 8.1
Drug effect by baseline IOP at week 12.

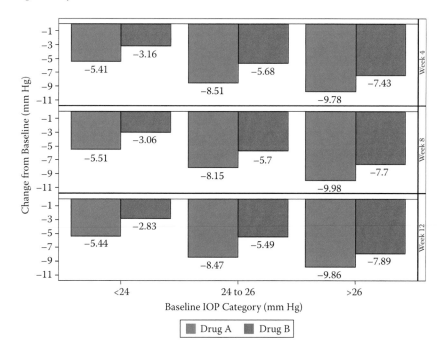

FIGURE 8.2
Drug effect by baseline IOP at weeks 4, 8, and 12.

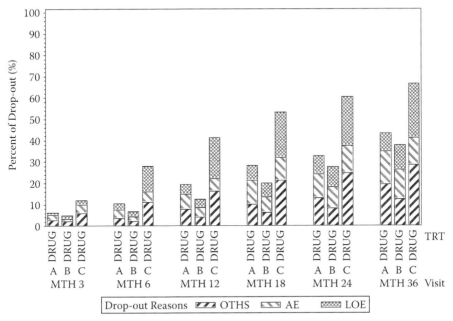

Sample Size: DRUG A (N = 150); DRUG B (N = 165); DRUG C (N = 145)

FIGURE 8.3
Stacked bar charts for patient dropout reasons by treatment and visit.

8.2.2 Example 2: Stacked Bar Charts for Patient Drop-out Reasons by Treatment and Visit

Stacked or subgroup bar charts showing the distribution of different subgroups (stacked groups) among the cluster groups is a good visualization tool. These types of bar charts can be done in the GPLOT and SGPANEL procedures.

Figure 8.3 is a stacked bar chart displaying the distribution of patients' dropout reasons by treatment, clustered by visits. Three dropout reasons—adverse event (AE), lack of efficacy (LOE), and others (OTHS)—are shown. From the figure, we can see that the dropout rates increased as time increased from month 3 to month 36. Drug C had the highest dropout rates among the three drugs at all visits, and the most dropouts fall into the OTHS category.

8.3 Producing the Sample Figures

8.3.1 Data Structure

Table 8.1 displays the data structure for the final dataset (CHG_SUM) that was used to produce the first two bar charts. The mean IOP reductions by

TABLE 8.1

Data Prepared to Produce Bar Charts in
Figures 8.1 and 8.2

visit	trtnum	bsl_c	mn_chg	n
Week 4	Drug A	< 24	−5.41	163
Week 4	Drug A	24 to 26	−8.51	85
Week 4	Drug A	> 26	−9.78	12
Week 4	Drug B	< 24	−3.16	152
Week 4	Drug B	24 to 26	−5.68	67
Week 4	Drug B	> 26	−7.43	21
Week 8	Drug A	< 24	−5.51	163
Week 8	Drug A	24 to 26	−8.15	85
Week 8	Drug A	> 26	−9.98	12
Week 8	Drug B	< 24	−3.06	152
Week 8	Drug B	24 to 26	−5.7	67
Week 8	Drug B	> 26	−7.7	21
Week 12	Drug A	< 24	−5.44	163
Week 12	Drug A	24 to 26	−8.47	85
Week 12	Drug A	> 26	−9.86	12
Week 12	Drug B	< 24	−2.83	152
Week 12	Drug B	24 to 26	−5.49	67
Week 12	Drug B	> 26	−7.89	21

treatment group (Drug A and Drug B) at each baseline IOP category are derived from a simulated dataset (IOP). Table 8.2 displays the data from the two visits that was used to produce the stacked bar chart in Figure 8.3.

8.3.2 Notes to SAS Programs

The two SAS programs that were used to produce the three sample bar charts using both PROC GPLOT and PROC SGPANEL are provided in the Appendix (Section 8.6). The following are the main sections and features of the programs.

8.3.2.1 Dataset Simulation (1ˢᵗ Program)

An IOP dataset is simulated to include 500 subjects, with each subject randomly assigned to either Drug A or Drug B at a 1:1 ratio. IOP values are assigned based on the treatment group and visit/hours with the two treatment groups having the same IOP values at baseline and Drug B having higher IOP values at post-baseline visits.

8.3.2.2 Data Analyses and Manipulation (1ˢᵗ Program)

Calculate the mean diurnal IOP, and the mean IOP at hours 0, 2, and 8 at each visit. The baseline IOP category (<24, 24 to 26, and >26 mm Hg) is calculated

TABLE 8.2

Part of the Data Used to Produce the
Stacked Bar Charts in Figure 8.3

visit	trt	reason	PCT
MTH 3	DRUG A	OTHS	2.84
MTH 3	DRUG A	AE	2.23
MTH 3	DRUG A	LOE	1
MTH 3	DRUG B	OTHS	2.2
MTH 3	DRUG B	AE	1
MTH 3	DRUG B	LOE	1.6
MTH 3	DRUG C	OTHS	5.85
MTH 3	DRUG C	AE	4.03
MTH 3	DRUG C	LOE	2.21
MTH 6	DRUG A	OTHS	3.45
MTH 6	DRUG A	AE	4.07
MTH 6	DRUG A	LOE	2.84
MTH 6	DRUG B	OTHS	2.2
MTH 6	DRUG B	AE	2.2
MTH 6	DRUG B	LOE	2.2
MTH 6	DRUG C	OTHS	11.3
MTH 6	DRUG C	AE	4.64
MTH 6	DRUG C	LOE	11.91

for each subject together with the change of mean diurnal IOP at each post-baseline visit.

8.3.2.3 Producing Bar Charts Using GPLOT and GREPLAY

The GPLOT and GREPLAY procedures were used to produce Figures 8.1 and 8.2.

- PROC GCHART is used to produce bar charts comparing the drug effect in IOP reduction by baseline IOP category at weeks 4, 8, and 12. Figure 8.1 shows the effect at week 12.

```
proc gchart data = chg_sum gout = work.bchart;
   where visit = 4;
   vbar trtnum/sumvar = mn_chg MIDPOINTS = (1, 2) group =
bsl_c width = 10
      space = 1.5 gspace = 3 legend = legend1 maxis = axis1
raxis = axis2
      gaxis = axis3 outside = sum name = "WK12";
   format bsl_c bsldf.;
run;
```

- Similar GOPTIONS for GPLOT can be used for GCHART. The following are some noteworthy features on GCHART.
  - **Axes:** Unlike GPLOT, GCHART has three axes with MAXIS for Midpoint, RAXIS for Response, and GAXIS for Group. In example Figure 8.1, the midpoint axis is for treatment (drugs), the response axis is for IOP change, and the group axis is for the baseline IOP category.
  - **Bar width and spaces:** WIDTH is used to specify bar width, SPACE specifies the space between bars, and GSPACE specifies the space between groups.
  - **Subgroup (stacked) bar charts:** SUBGROP is used to specify subgroup (stacked) variables. In Figure 8.3, the stacked group is for dropout reason and it is produced using *subgroup = reason*.
  - **Distinguishing bars:** To distinguish the bars, we can choose different bar patterns together with colors. In SAS, there are a total of 17 patterns from which to choose, with 5 for left-slash (L1 to L5), 5 for right-slash (R1 to R5), 5 for crosses (X1 to X5), 1 for empty (E), and 1 for solid (S).
- The graph output category is specified using the GOUT option. The three bar charts of weeks 4, 8, and 12 are saved in the WORK category in the names WK4, WK8, and WK12 using the NAME = option in the VBAR statement.
- The three bar charts of weeks 4, 8, and 12 saved in the output category are processed and positioned in the same page using the GREPLAY procedure in Figure 8.2. The SAS-provided template (V3, with 3 boxes stacked vertically) is used to combine the 3 bar charts on one page.

```
proc greplay igout = work.bchart;
  tc = sashelp.templt;
  template = v3;
  ** Replay into the chosen template;
  treplay 1:WK4 2:WK8 3:WK12;
run;
quit;
```

8.3.2.4 Producing Bar Charts Using SGPLOT and SGPANEL

- PROC SGPLOT is used to produce Figure 8.1 with drug effects by baseline IOP category at week 12.
  - VBARPARM is a new statement in SGPLOT (available in SAS 9.3) that is used to create a vertical bar chart based on a presumma-rized response value for each unique value of the category variable.

```
ods listing style = MONOCHROMEPRINTER gpath = "&outloc.";
ods graphics/reset = all width = 8in height = 6in
noborder OUTPUTFMT = emf imagename = "Fig_8_1";
title1 &GRAPHTTL "Figure 8.1 Treatment Effect by Baseline
IOP at Week 12";
footnote1 &GRAPHFOOT "&pgmpth.";
proc sgplot data = chg_sum;
   where visit = 4;
   vbarparm category = bsl_c response = mn_chg/
   group = trtnum dataskin = pressed datalabel;
   keylegend/location = outside position = bottom;
   xaxis display = (nolabel) label = "Baseline Average Eye
Mean Diurnal
      IOP (mm Hg)";
   yaxis VALUES = (-11 to 0 by 1) label = "Change in Average
Eye Mean
      Diurnal IOP (mm Hg)";
   format bsl_c bsldf.;
run;
quit;
```

- The fill patterns to distinguish the grouped bar charts can be specified by using the STYLE = option in an ODS Listing statement. The default styles include JOURNAL2, JOURNAL3 (uses gray and the fill pattern), and MONOCHROMEPRINTER.

- The DATASKIN option is used to specify a special effect to be used on all filled bars. Available options include NONE | CRISP | GLOSS | MATTE | PRESSED | SHEEN.

- PROC SGPANEL is used to produce paneled bar charts by visits. The following are some noteworthy features on SGPANEL.

  - SGPANEL in ODS Graphics is a versatile tool for producing multipanel classification panels. The sample bar charts are produced by using the VBARPARM plot statement.

  - Panels: PANELBY is used to specify the paneling structure of the graph. Figure 8.3 is a paneled bar chart by visit.

  - Axes: COLAXIS and ROWAXIS are used to specify the column and row axes.

- Style = MONOCHROMEPRINTER is specified in an ODS RTF statement to save figures in monochrome and fill patterns for bars. If STYLE is not specified, the figures would be in color and without fill patterns.

8.3.2.5 Bar Charts with Stacked and Cluster Groups (*2nd Program*)

Figure 8.3 is a bar chart with stacked (dropout reason) and cluster (treatment and visit) groups. This bar chart can be done in GCHART using the VBAR, SUBGROUP, and GROUP options. SGPLOT can easily produce the bar charts with one or two classification groups. The bar charts can be stacked or side by side, but SGPLOT cannot produce stacked bar charts with cluster groups that have three levels of categorization (Matange, 2013). The good news is that this can be done easily using the SGPANEL procedure.

```
legend1 label = (height = 3 'Drop-out Reasons');
pattern1 color = black value = r4;
pattern2 color = red value = l4;
pattern3 color = blue value = x2;
axis1 label = (height = 2 "TRT" justify = center) value =
(height = 2);
axis2 order = (0 to 100 by 10) label = (angle = 90 height
= 2.5 "Percent of Drop-out (%)");
proc gchart data = dropout;
   vbar trt/sumvar = pct subgroup = reason group = visit
      MIDPOINTS = (1, 2, 3) width = 8 space = 0 gspace = 6
      maxis = axis1 raxis = axis2 legend = legend1;
   format visit vstdf. reason rsndf. trt trtdf.;
   label visit = "Visit";
run;
quit;

proc sgpanel data = dropout;
   panelby visit/layout = columnlattice novarname noborder
      colheaderpos = bottom columns = 6;
   vbar trt/response = pct group = reason dataskin =
gloss;
      keylegend/position = bottom noborder;
   colaxis display = (nolabel);
   rowaxis VALUES = (0 to 100 by 10) grid label = "Percent
of Drop-out (%)";
run;
```

8.4 Summary and Discussion

Bar charts can be produced using the PROC GCHART procedure in SAS/GRAPH and the PROC SGPLOT procedure in ODS Graphics. The main features, including the pros and cons of using PROC GCHART and PROC SGPPLOT in producing the bar charts, are summarized in Table 8.3.

TABLE 8.3

Comparing PROC GCHART and PROC SGPLOT in Producing Bar Charts

Features	GCHART	SGPLOT
Axes	3 axes: MAXIS for midpoint; RAXIS for response and GAXIS for group variable	XAXIS and YAXIS plot statements
Bar Charts	Using VBAR, HBAR	Using VBARPARM, HBARPARM, VBAR, HBAR
Summary, Categorization and Grouping Variables	VBAR = the direct comparing variable, i.e., trtnum SUMVAR = for response summary variable GROUP = for group category SUBGROUP = for subgroup category	RESPONSE = for response summary variable; CATEGORY = for categories (equivalent to GROUP in GCHART); GROUP = for comparing groups. No SUBGROUP option
Fill Patterns/Styles	17 default patterns: 5 for left-slash (L1 to L5), 5 for right-slash (R1 to R5), 5 for crosses (X1 to X5), 1 for empty (E), and 1 for solid (S).	Only 3 included default styles: JOURNAL2, JOURNAL3, and MONOCHROMEPRINTER. Other custom styles can be specified using PROC Template.
Multiple Bar Charts in One Page	Using PROC GREPLAY: must save each individual bar chart in a category first.	Using PROC SGPANEL with PANEL by statement; no need to save individual bar chart.
Pros	More available default fill patterns. Can produce stacked bar charts with cluster groups that has 3 levels of classification.	Easier axes and summary, categorization, and group variables.
Cons	The 3 axes can be confusing and labels are positioned to the right end of the axes.	Fewer default fill patterns or styles. Cannot produce stacked bar charts with cluster groups that have 3 classification levels.

For a detailed description of PROC GCHART and other applications, please see SAS/GRAPH 9.3 reference manuals. Watts also has a good paper on the basics of PROC GCHART (Watts, 2007).

8.5 References

Matange, S. 2013. "Bar Charts with Stacked and Cluster Groups." SAS Institute Inc. blogs, September 7, http://blogs.sas.com/content/graphicallyspeaking/2013/09/07/bar-charts-with-stacked-and-cluster-groups/.

SAS Institute, Inc. "The GCHART Procedure." In *SAS/GRAPH® 9.2: Reference*, 2nd ed. Cary, NC: SAS Institute Inc., http://support.sas.com/documentation/cdl/en/graphref/63022/HTML/default/viewer.htm#a000723580.htm.

Watts, P. 2007. "Charting the Basics with PROC GCHART." In *NESUG 2007 Proceedings*,
    http://www.nesug.org/proceedings/nesug07/ff/ff17.pdf.
Wikipedia. "Bar Chart," http://en.wikipedia.org/wiki/Bar_chart.

8.6 Appendix: SAS Programs for Producing the Sample Figures

8.6.1 Bar Charts

```
***************************************************************;
* Program Name: Chapter 8 Bar Charts.sas                     *;
* Descriptions: Producing the following sample figures in    *;
* chapter 8                                                  *;
* - Figure 8.1 Treatment Effect by Baseline IOP at Week 12   *;
* - Figure 8.2 Treatment Effect by Baseline IOP and Visit    *;
* Lattice by Treatment and Visit                             *;
***************************************************************;
options mprint symbolgen nodate nonumber validvarname = v7
orientation = landscape;
%let pgmname = Chapter 8 Bar Charts.sas;
%let pgmloc = C:\SASBook\SAS Programs;
%let outloc = C:\SASBook\Sample Figures\Chapter 8;
%let pgmpth = &pgmloc.\&pgmname. &sysdate9. &systime. SAS
V&sysver.;

** Set-up macro variables for data simulation;
%LET SD = 3.5;
%let sitenum = 10;
%let seed = 08;
%let subjnum = 500;

proc format;
  value trtdf
    1 = 'Drug A'
    2 = 'Drug B'
    OTHER = ' ';
  value visdf
    1 = 'Baseline'
    2 = 'Week 4'
    3 = 'Week 8'
    4 = 'Week 12';
  value hrdf
    1 = 'Hour 0'
    2 = 'Hour 2'
    3 = 'Hour 8';
  value bsldf
    1 = '< 24'
    2 = '24 to 26'
```

```
    3 = '> 26'
    OTHER = ' ';
run;

** Generate the required number of subjects and randomly
assign to 2 treatment groups;
data subj;
  do i = 1 to &subjnum.;
    subjid = 1000+ i;
    if ranuni (&seed.) < 0.5 then trtnum = 1;
    else trtnum = 2;
    output;
  end;
  drop i;
run;

** Set up the IOP Values based on the trt assignment and
visits/timepoints;
data iop;
  set subj;
  do i = 1 to 4; ** 4 visits;
    do j = 1 to 3; ** 3 timepoints/visit;
      visit = i;
      hour = j;
      if i = 1 then do; ** Baseline;
        if j = 1 then iop = round((RANNOR(&seed.)* &SD.
+ 25),.1); ** Hour 0;
        if j = 2 then iop = round((RANNOR(&seed.)* &SD.
+ 23),.1); ** Hour 2;
        if j = 3 then iop = round((RANNOR(&seed.)* &SD.
+ 22),.1); ** Hour 8;
      end;
      else if i > 1 and trtnum = 1 then do; ** Post-baseline:
drug A;
        if j = 1 then iop = round((RANNOR(&seed.)* &SD.
+ 17.5),.1); ** Hour 0;
        if j = 2 then iop = round((RANNOR(&seed.)* &SD.
+ 16.5),.1); ** Hour 2;
        if j = 3 then iop = round((RANNOR(&seed.)* &SD.
+ 16.2),.1); ** Hour 8;
      end;
      else if i > 1 and trtnum = 2 then do; ** Post-baseline:
drug B;
        if j = 1 then iop = round((RANNOR(&seed.)* &SD.
+ 20),.1); ** Hour 0;
        if j = 2 then iop = round((RANNOR(&seed.)* &SD.
+ 19),.1); ** Hour 2;
        if j = 3 then iop = round((RANNOR(&seed.)* &SD.
+ 18.7),.1); ** Hour 8;
      end;
      output;
    end;
```

```
  end;
  drop i j;
  format visit visdf. hour hrdf. trtnum trtdf.;
run;

proc sort data = iop;
  by subjid visit hour;
run;

** Mean diurnal IOP at each visit: mean of hours 0, 2 and 8
average eye IOP at baseline;
proc transpose data = iop out = iop_t;
  by subjid trtnum visit;
  var iop;
  id hour;
run;

data iop_MnDiur;
  set iop_t;
  iop_MnDiur = round (mean (Hour_0, Hour_2, Hour_8),.01);
  drop _NAME_;
run;

** baseline mean ddiurnal IOP and category;
data mndiur_bsl;
  set iop_MnDiur;
  where visit = 1;
  if iop_MnDiur < 24 then bsl_c = 1;
  if 24 < = iop_MnDiur < = 26 then bsl_c = 2;
  if iop_MnDiur > 26 then bsl_c = 3;
  rename iop_MnDiur = mndiur_bsl;
  drop visit;
run;

** change from baseline;
data mndiur_chg;
  merge iop_MnDiur (where = (visit > 1)) mndiur_bsl;
  by subjid;
  chg = iop_MnDiur - mndiur_bsl;
  chg_pct = (chg/mndiur_bsl)*100;
run;

** Summary for change from baseline;
proc means data = mndiur_chg noprint;
  class visit trtnum bsl_c;
  var chg;
  output out = chg_stat mean = mn_chg n = n;
run;
```

```
data chg_sum;
   set chg_stat;
   where visit > 1 and trtnum ne. and bsl_c ne.;
   mn_chg = round (mn_chg,.01);
   format bsl_c bsldf.;
   drop _FREQ_ _TYPE_;
run;

** save the number of patients at each baseline category and
treatment;
** to macro variables to be used later;
data _NULL_;
   set chg_stat;
   where visit = 4 and bsl_c ne.;
   if trtnum = 1 then do;
      call symputx('N_DA_C'||left(put(bsl_c,best.)), put(N, 3.0));
   end;
   if trtnum = 2 then do;
      call symputx('N_DB_C'||left(put(bsl_c,best.)), put(N, 3.0));
   end;
run;

** Macro to produce bar charts;
%LET FONTNAME = Times;
%LET DRIVER = PSCOLOR;%LEt EXT = PS;
goptions
   reset = all
   GUNIT = PCT
   rotate = landscape
   gsfmode = replace
   gsfname = GSASFILE
   device = &DRIVER
   lfactor = 1
   hsize = 8 in
   horigin = 0 in
   vsize = 6 in
   vorigin = 6 in
   ftext = "&FONTNAME"
   htext = 10pt
   ftitle = "&FONTNAME"
   htitle = 10pt
;

pattern color = black value = x3;
axis1 major = none minor = none order = (1 to 2 by 1)
   label = (h = 2.5 font = "&FONTNAME" "Treatment");
axis2 order = (-11 to 0 by 1) minor = none label = (a = 90 r =
0 h = 2.5
```

```
  font = "&fontname" "Change from Baseline at Week 4 (mm Hg)");
axis3 value = (height = 2.5) label = (height = 2.5 "BSL IOP
(mm Hg)" justify = left);

title1 "Treatment Effect by Baseline IOP at Week 4";
footnote1 Raindrop_SGPLOT"&pgmpth.";
FILENAME GSASFILE "&OUTLOC.\Week 4.&EXT.";
proc gchart data = chg_sum gout = work.bchart;
  where visit = 2;
  vbar trtnum/sumvar = mn_chg MIDPOINTS = (1, 2) group = bsl_c
width = 10 space = 1.5
    gspace = 3 legend = legend1 maxis = axis1 raxis = axis2
gaxis = axis3 outside = sum name = "WK4";
  format bsl_c bsldf.;
run;

axis2 order = (-11 to 0 by 1) minor = none label = (a = 90 r = 0
h = 2.5
  font = "&fontname" "Change from Baseline at Week 8 (mm Hg)");
title1 "Treatment Effect by Baseline IOP at Week 8";
FILENAME GSASFILE "&OUTLOC.\Week 8.&EXT.";
proc gchart data = chg_sum gout = work.bchart;
  where visit = 3;
  vbar trtnum/sumvar = mn_chg MIDPOINTS = (1, 2) group = bsl_c
width = 10 space = 1.5
  gspace = 3 legend = legend1 maxis = axis1 raxis = axis2
gaxis = axis3 outside = sum name = "WK8";
  format bsl_c bsldf.;
run;

axis2 order = (-11 to 0 by 1) minor = none label = (a = 90 r = 0
h = 2.5
  font = "&fontname" "Change from Baseline at Week 12 (mm Hg)");
title1 "Figure 8.1 Treatment Effect by Baseline IOP at Week 12";
FILENAME GSASFILE "&OUTLOC.\Week 12.&EXT.";
proc gchart data = chg_sum gout = work.bchart;
  where visit = 4;
  vbar trtnum/sumvar = mn_chg MIDPOINTS = (1, 2) group = bsl_c
width = 10 space = 1.5
  gspace = 3 legend = legend1 maxis = axis1 raxis = axis2
gaxis = axis3 outside = sum name = "WK12";
  format bsl_c bsldf.;
run;

** Put the 3 figures in one page using PROC GREPLAY;
title1 "Figure 8.2 Treatment Effect by Baseline IOP at each
Visit";
```

```
footnote1 Raindrop_SGPLOT"&pgmpth.";
FILENAME GSASFILE "&OUTLOC.\Figure 8.2.&EXT.";
proc greplay igout = work.bchart;
   tc = sashelp.templt;
   template = v3;
   ** Replay into the chosen template;
   treplay 1:WK4 2:WK8 3:WK12;
run;
quit;

****************************************************************;
* Produce Bar charts in SGPLOT and SGPANEL                   *;
****************************************************************;
%LET OUTPUTFMT = PS;
ods listing style = MONOCHROMEPRINTER gpath = "&outloc.";
ods graphics/reset = all width = 8in height = 6in noborder
OUTPUTFMT = &OUTPUTFMT. imagename = "Fig_8_1";
title1 "Figure 8.1 Treatment Effect by Baseline IOP at Week 12";
footnote1 Raindrop_SGPLOT"&pgmpth.";
proc sgplot data = chg_sum;
   where visit = 4;
   vbarparm category = bsl_c response = mn_chg/
   group = trtnum dataskin = none datalabel;
   keylegend/position = bottom noborder;
   xaxis label = "Baseline IOP Category (mm Hg)";
   yaxis VALUES = (-11 to 0 by 1) label = "Change from Baseline
(mm Hg)";
   format bsl_c bsldf.;
run;
quit;

ods graphics/reset = all width = 8in height = 6in noborder
OUTPUTFMT = &OUTPUTFMT. imagename = "Fig_8_2";
   title1 "Figure 8.2 Treatment Effect by Baseline IOP and Visit";
   title2 "Lattice by Viist";
   footnote1 Raindrop_SGPLOT"&pgmpth.";
proc sgpanel data = chg_sum;
   where visit > 1;
   panelby visit/novarname layout = ROWLATTICE onepanel;
   vbarparm category = bsl_c response = mn_chg/
   group = trtnum dataskin = none datalabel;
   keylegend/position = bottom noborder;
   colaxis label = "Baseline IOP Category (mm Hg)";
   rowaxis VALUES = (-11 to 0 by 1) label = "Change from Baseline
(mm Hg)";
   format bsl_c bsldf.;
run;
quit;
```

8.6.2 Stacked Bar Charts

```
*****************************************************************;
* Program Name: Chapter 8 Stacked Bar Charts.sas              *;
* Descriptions: Producing the following sample figures in     *;
* Chapter 8                                                   *;
* - Figure 8.3 Stacked Bar Charts for the Distribution        *;
* of Patient                                                  *;
* Drop-out Reasons by Treatment Group                         *;
*****************************************************************;
options mprint symbolgen nodate nonumber validvarname = v7
orientation = landscape;
%let pgmname = Chapter 8 Stacked Bar Charts.sas;
%let pgmloc = C:\SASBook\SAS Programs;
%let outloc = C:\SASBook\Sample Figures\Chapter 8;
%let pgmpth = &pgmloc.\&pgmname. &sysdate9. &systime. SAS
V&sysver.;

proc format;
  value vstdf
     1 = "MTH 3"
     2 = "MTH 6"
     3 = "MTH 12"
     4 = "MTH 18"
     5 = "MTH 24"
     6 = "MTH 36"
     OTHER = " ";
  value trtdf
     1 = "DRUG A"
     2 = "DRUG B"
     3 = "DRUG C"
     OTHER = " ";
  value rsndf
     1 = "OTHS"
     2 = "AE"
     3 = "LOE";
run;

%let N_G1 = 150;%let N_G2 = 165;%let N_G3 = 145;

data dropout;
  input visit trt reason PCT;
  format visit vstdf. trt trtdf. reason rsndf.;
  datalines;
1 1 1 2.84
1 1 2 2.23
1 1 3 1.00
1 2 1 2.20
1 2 2 1.00
1 2 3 1.60
```

```
1 3 1 5.85
1 3 2 4.03
1 3 3 2.21
2 1 1 3.45
2 1 2 4.07
2 1 3 2.84
2 2 1 2.20
2 2 2 2.20
2 2 3 2.20
2 3 1 11.30
2 3 2 4.64
2 3 3 11.91
3 1 1 7.75
3 1 2 7.13
3 1 3 4.68
3 2 1 4.01
3 2 2 4.61
3 2 3 4.01
3 3 1 16.15
3 3 2 5.85
3 3 3 19.18
4 1 1 10.20
4 1 2 11.20
4 1 3 6.91
4 2 1 6.22
4 2 2 7.42
4 2 3 6.22
4 3 1 21.39
4 3 2 10.48
4 3 3 21.39
5 1 1 13.04
5 1 2 11.20
5 1 3 8.52
5 2 1 8.42
5 2 2 9.63
5 2 3 9.63
5 3 1 24.82
5 3 2 12.70
5 3 3 23.00
6 1 1 19.56
6 1 2 15.27
6 1 3 8.52
6 2 1 12.64
6 2 2 13.84
6 2 3 11.43
6 3 1 28.45
6 3 2 12.70
6 3 3 25.42
;
run;
```

```
%LET FONTNAME = Times;
%LET DRIVER = PSCOLOR;%LEt EXT = PS;
goptions
   reset   = all
   GUNIT   = PCT
   rotate  = landscape
   gsfmode = replace
   gsfname = GSASFILE
   device  = &DRIVER
   lfactor = 1
   hsize   = 8 in
   horigin = 0 in
   vsize   = 6 in
   vorigin = 0 in
   ftext   = "&FONTNAME"
   htext   = 10pt
   ftitle  = "&FONTNAME"
   htitle  = 10pt
;

legend1 label = (height = 3 'Drop-out Reasons');
**pattern1 color = black value = e;
pattern1 color = black value = r4;
pattern2 color = red value = l4;
pattern3 color = blue value = x2;

axis1 label = (height = 2 "TRT" justify = center) value =
(height = 2);
axis2 order = (0 to 100 by 10) label = (angle = 90 height = 2.5
"Percent of Drop-out (%)");

title1 "Figure 8.3. Stacked Bar Charts for the Distribution of
Patient Drop-out Reasons by Treatment Group";
footnote1 "Sample Size in ITT: DRUG A (N =%trim(&N_G1.)); DRUG B
(N =%trim(&N_G2.)); DRUG C (N =%trim(&N_G3.))";
footnote2 "&pgmpth.";
FILENAME GSASFILE "&OUTLOC./&fignum.&EXT.";

proc gchart data = dropout;
   vbar trt/sumvar = pct subgroup = reason group = visit
      MIDPOINTS = (1, 2, 3) width = 8 space = 0 gspace = 6
      maxis = axis1 raxis = axis2 legend = legend1;
   format visit vstdf. reason rsndf. trt trtdf.;
   label visit = "Visit";
run;
quit;

%LET OUTPUTFMT = ps;
ods listing gpath = "&outloc.";
```

```
ods graphics/reset = all width = 8in height = 6in noborder
OUTPUTFMT = &OUTPUTFMT. imagename = "Fig_8_3";
title1 "Figure 8.3. Stacked Bar Charts for the Distribution of
Patient Drop-out Reasons by Treatment Group";
footnote1 "&pgmpth.";
proc sgpanel data = dropout;
  panelby visit/
  layout = columnlattice novarname noborder colheaderpos =
bottom columns = 6;
  vbar trt/response = pct group = reason dataskin = gloss;
  keylegend/position = bottom noborder;
  colaxis display = (nolabel);
  rowaxis VALUES = (0 to 100 by 10) grid label = "Percent of
Drop-out (%)";
run;
```

9

Box Plots

9.1 Introduction

Unlike line plots and bar charts, which mainly provide data visualization for mean or median values, a box plot "summarizes the data and indicates the median, upper, and lower quartiles and minimum and maximum values" (SAS Institute Inc., 2012a). A box plot provides a quick visual summary that easily shows center, spread, range, and any outliers. This chapter introduces and illustrates the production of box plots using both GPLOT and SGPLOT procedures.

9.2 Application Examples

To illustrate the application and production of box plots, two sample figures are presented in the chapter based on clinical research in the glaucoma therapeutic area. The same simulated data used in Chapter 8 for bar charts are used to produce the box plots in this chapter.

9.2.1 Example: Box Plots for Drug Effects on IOP Reduction by Baseline IOP Categories

It is commonly considered that the drug's effect on a patient's IOP reduction is influenced by the patient's baseline or initial IOP values. Box plots of the drug's effects on IOP reduction by different baseline IOP categories (< 24, 24 to 26, and > 26 mm Hg) for Drug A and Drug B are shown in Figures 9.1 and 9.2. Figure 9.1 displays the drug effects at week 12 by baseline IOP categories while Figure 9.2 displays the drug effects by baseline IOP category for all the three post-baseline visits. Figure 9.2 is a classification box plot and is produced using GREPLAY in SAS/GRAPH and SGPANEL in ODS Graphics.

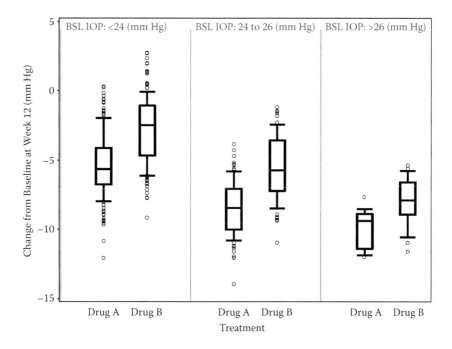

FIGURE 9.1

Treatment effect by baseline IOP at week 12 (produced in GPLOT). Box plots display median, 25th, and 75th percentiles with whiskers extended to 10th and 90th percentiles. Any values outside those values are considered as outliers.

9.3 Producing the Sample Figures

9.3.1 Data Structure

The same simulated dataset is used for box plots in this chapter as for bar chats in Chapter 8. Part of the simulated dataset for a few subjects containing variables for IOP change from baseline, baseline category, and so on is shown in Table 9.1.

9.3.2 Notes to SAS Programs

Detailed SAS programs to produce the sample plots using both the GPLOT in SAS/GRAPH and the SGPLOT in ODS Graphics are provided in the Appendix (Section 9.7). The programs consist of the following main sections.

9.3.2.1 Dataset Simulation

An IOP dataset is simulated to include 500 subjects, with each subject randomly assigned to either Drug A or Drug B at a 1:1 ratio. IOP values are assigned based on the treatment group assignment and visits. The two

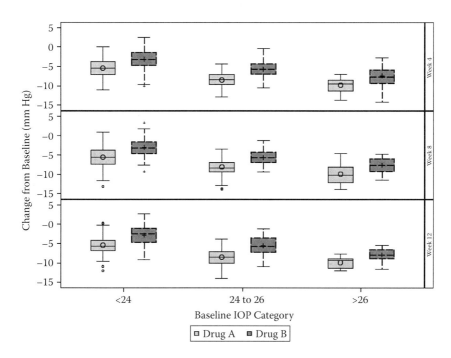

FIGURE 9.2

Treatment effect by baseline IOP at week 12 (produced in SGPLOT). Box plots display median, 25th, and 75th percentiles with whiskers extended to 1.5* IQR from the mean value (by default).

TABLE 9.1

Part of the Dataset Used to Produce Box Plots

subjid	trtnum	visit	iop_MnDiur	mndiur_bsl	bsl_c	chg	index
1003	Drug A	Week 4	15.73	22.13	1	−6.4	1
1004	Drug A	Week 4	14.67	22.17	1	−7.5	1
1010	Drug B	Week 4	17	23.83	1	−6.83	2
1011	Drug B	Week 4	15.17	20.3	1	−5.13	2
1001	Drug A	Week 4	15.63	25.33	2	−9.7	4

treatment groups are designed to have the same IOP values at baseline and subjects in the Drug B group have higher IOP values at post-baseline visits.

9.3.2.2 Data Analyses and Manipulation

The mean diurnal IOP and the mean of IOP values at hours 0, 2, and 8 at each visit are calculated. The baseline IOP category (< 24, 24 to 26, and > 26 mm Hg) is calculated for each subject together with the change from baseline in mean diurnal IOP at each post-baseline visit. Part of the dataset is shown in Table 9.1.

A variable *index* is created to represent the treatment groups in the three baseline IOP categories and values of 1, 2, 4, 5, 7, and 8 are assigned. The values of 3 and 6 are not used in order to leave space for the reference lines. Together with the format statements, the reference labels, the treatment, and the baseline IOP categories can be displayed in the plot (Figure 9.1).

9.3.2.3 Producing the Sample Figures Using GPLOT and GREPLAY

- PROC GPLOT is used to produce box plots comparing the drug effect in IOP reduction by baseline IOP category at weeks 4, 8, and 12. The box plot is produced by specifying the INTEPOL = BOXT10 option in the SYMBOL statement. Figure 9.1 displays the effect at week 12.

```
SYMBOL H = 2 CV = Black CO = Black INTERPOL = BOXT10 W = 3
bwidth = 5 VALUE = CIRCLE;
title1 "Figure 9.1 Treatment Effect by Baseline IOP at
Week 12";
FILENAME GSASFILE "&OUTLOC.\Week 12.&EXT.";
proc gplot data = mndiur_chg2 gout = work.bplot;
   where visit = 4;
   plot chg * index/vaxis = axis2 haxis = axis1 href = 0
href = 3 href = 6 name = "WK12";
   format index trtdf.;
run;
quit;
```

- A graph output category is specified using GOUT option. The three box plots of weeks 4, 8, and 12 are saved in the specified category with the names WK4, WK8, and WK12, respectively, using the NAME = option in the PLOT statement.
- The three box plots of weeks 4, 8, and 12 are saved in the output category in the WORK directory and are processed and positioned in the same page using the GREPLAY procedure in Figure 9.2. The SAS-provided template (V3, with 3 boxes stacked vertically) is used to combine the 3 box plots in one page.

```
proc greplay igout = work.bplot;
   tc = sashelp.templt;
   template = v3;
   ** Replay into the chosen template;
   treplay 1:WK4 2:WK8 3:WK12;
run;
quit;
```

9.3.2.4 Producing the Sample Figures in SGPLOT and SGPANEL

- PROC SGPLOT is used to produce Figure 9.1 with drug effects by baseline IOP category at week 12.

```
ods listing style = JOURNAL gpath = "&outloc.";
ods graphics/reset = all width = 8in height = 6in
noborder OUTPUTFMT = &OUTPUTFMT. imagename = "Fig_9_1";
title1 &GRAPHTTL "Figure 9.1 Treatment Effect by Baseline
IOP at Week 12";
footnote1 &GRAPHFOOT "&pgmpth.";
proc sgplot data = mndiur_chg2;
  where visit = 4;
  vbox chg/category = bsl_c group = trtnum SPREAD;
  keylegend/position = bottom noborder;
  xaxis VALUES = (0 to 8.5 by.5) label = "Treatment by
Baseline IOP Category";
  yaxis VALUES = (-15 to 5 by 5) label = "Change from
Baseline (mm Hg)";
  format bsl_c bsldf.;
run;
quit;
```

- The VBOX statement is used to produce vertical box plots. Horizontal box plots can be produced using the HBOX statement.

- The fill patterns to distinguish the grouped box plots can be specified by using the STYLE = option in the ODS Listing statement. The default styles include JOURNAL2, JOURNAL3 (uses gray and the fill pattern), and MONOCHROMEPRINTER.

- The DATASKIN option is used to specify a special effect to be used on all filled bars. Available options include NONE | CRISP | GLOSS | MATTE | PRESSED | SHEEN.

- PROC SGPANEL is used to produce paneled box plots by visits. The following are some noteworthy features of SGPANEL.

  - SGPANEL in ODS Graphics is a versatile tool to produce multi-panel classification panels. The sample box plots are produced by using VBOX plot statement.

  - Panels: PANELBY is used to specify the paneling structure of the graph. Figure 9.3 is a paneled box plot by visit.

  - Axes: COLAXIS and ROWAXIS are used to specify the column and row axes.

```
ods graphics/reset = all width = 8in height = 6in noborder
OUTPUTFMT = &OUTPUTFMT. imagename = "Fig_9_2";
title1 "Figure 9.2 Treatment Effect by Baseline IOP and
Visit";
title2 "Lattice by Visit";
footnote1 "&pgmpth.";
proc sgpanel data = mndiur_chg2;
   panelby visit/novarname layout = ROWLATTICE onepanel;
   vbox chg/category = bsl_c group = trtnum;
   keylegend/position = bottom noborder;
   colaxis VALUES = (0 to 8.5 by.5) label = "Baseline IOP
Category";
   rowaxis VALUES = (-15 to 5 by 5) label = "Change from
Baseline (mm Hg)";
   format bsl_c bsldf.;
run;
quit;
```

9.3.2.5 Producing Custom Box Plots with Specified Whisker Length in SGPLOT

By default, the whiskers produced using the VBOX (or HBOX) statement in SGPLOT extend to 1.5* IQR (intraquartile range) range. Other than GPLOT, which can easily choose the length of the whisker by specifying the range in *INTERPOL = BOX<option(s)><00…25>* option, SGPLOT in SAS 9.3 does not support this option. Such an option has been added for SAS 9.4. With SAS 9.3, we can create a box plot with custom whisker lengths by computing the statistics for mean, median, q1, q3, p10, and p90 and using SGPLOT to plot them (Matange, 2013). If we want to add outliers (those outside the whiskers), more codes are needed.

```
ods html close;
ods listing style = htmlblue gpath = "&outloc." image_dpi
= 150;
ods graphics/reset = all width = 8in height = 6in noborder
imagename = "Fig_9_2b";
title1 "Figure 9.2 Treatment Effect by Baseline IOP at
Week 12";
footnote1 "&pgmpth.";
proc sgplot data = chg_sum nocycleattrs;
   where visit = 4;
   highlow x = bsl_c high = p90 low = p10/group = trtnum
groupdisplay = cluster
```

```
      clusterwidth = 0.7;
   highlow x = bsl_c high = q3 low = median/group = trtnum
type = bar
      groupdisplay = cluster grouporder = ascending
clusterwidth = 0.7 barwidth = 0.7 name = 'a';
   highlow x = bsl_c high = median low = q1/group = trtnum
type = bar
      groupdisplay = cluster grouporder = ascending
clusterwidth = 0.7 barwidth = 0.7;
   scatter x = bsl_c y = mean/group = trtnum groupdisplay
= cluster
   grouporder = ascending clusterwidth = 0.7 markerattrs =
(size = 10);
   keylegend 'a';
   xaxis label = "Baseline IOP Category";
   yaxis grid label = "Change from Baseline (mm Hg)";
   format bsl_c bsldf.;
run;
```

These programs are based on a blog article on custom box plots posted by Sanjay Matange from SAS (Matange, 2013). Here are the details of this program:

- The first high low plot of type = line (default) plots the whisker from P10 to P90.
- The second high low plot of type = bar draws the upper quartile.
- The third high low plot of type = bar draws the lower quartile.
- The scatter plot draws the mean marker.
- This graph looks very similar to the standard VBOX except for the whiskers and outliers.
- The HTMLBLUE style is used, and by default a PNG file with the specified resolution of 150 DPI is saved.

9.4 Understanding Box Plots

A box plot is composed of different parts, and in PROC GPLOT, the *INTERPOL = BOX<option(s)><00...25>* in the SYMBOL statement is used to specify the box-and-whisker plots (SAS Institute Inc., 2012a).

<option(s)>: can be one or more of the following values:

- F: fills the box with the color specified by CV = and outlines the box with the color specified by CO =.
- J: joins the median points of the boxes with a line.
- T: draws tops and bottoms on the whiskers.

<00…25>: Used to specify a percentile to control the length of the whiskers within the range 00 through 25. For example, 01 specifies 1st percentile low and 99th percentile high. By default, any box plot indicates the median, 25th, and 75th percentiles in the middle, low, and up end of the box. The INTERPOL = BOXT10 options used in the sample codes extend the box plots with the vertical lines (whiskers) drawn to the 10th and 90th percentiles. Any values more extreme than this are marked with a circle symbol.

In SGPLOT, by default, a box plot displays the 25th, 75th percentile interval, also known as the intraquartile range (IQR), the mean marker, median line, whiskers, and outliers (Figure 9.3). The whiskers that extend from each box

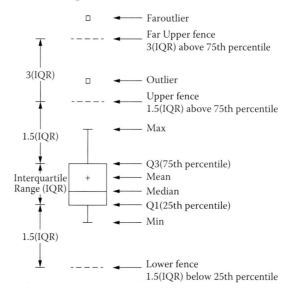

FIGURE 9.3
Parts of a box plot produced in SGPLOT. The bottom and top edges of the box indicate the intraquartile range (IQR). That is, the range of values between the first and third quartiles (the 25th and 75th percentiles). The marker inside the box indicates the mean value. The line inside the box indicates the median value. The elements that are outside the box depend on your options. By default, the whiskers that extend from each box indicate the range of values that are outside of the intraquartile range. However, the values are close enough not to be considered outliers (a distance less than or equal to 1.5*IQR). (Figure 6.5, Parts of a Box Plot, from SAS Institute Inc. 2012. *SAS® 9.3 ODS Graphics: Procedures Guide*, 3rd ed. Cary, NC: SAS Institute Inc.) Courtesy: SAS Institute, Inc. Cary, NC, USA.

indicate the range of values that are outside of the IQR, but are close enough not to be considered outliers (a distance less than or equal to 1.5*IQR) (SAS Institute Inc., 2012b). Any points that are a distance of more than 1.5*IQR from the box are considered to be outliers. By default, these points are indicated by markers. If you specify DATALABEL = option, then the outlier points have data labels. If you also specify the LABELFAR option, then only outliers that are 3*IQR from the box have data labels. If you specify the EXTREME option, then the whiskers indicate the entire range of values, including outliers (SAS Institute Inc., 2012b).

9.5 Summary and Discussion

Box plots can be produced using PROC GPLOT in SAS/GRAPH and PROC SGPLOT in ODS Graphics. The main features, including the pros and cons of using GPLOT and SGPPLOT to produce box plots, are summarized in Table 9.2.

In SGPLOT, the CATEGORY = option defines the variable to use for the x-axis, whereas the GROUP = option defines an auxiliary discrete variable whose values and graphical attributes are displayed in a legend (Wicklin, 2012). You can use the options to visualize the distribution of one response variable (e.g., treatment group) with respect to one or two other variables (e.g., baseline IOP category).

TABLE 9.2

Comparing GPLOT and SGPLOT in Producing Box Plots

Features	GPLOT	SGPLOT
Box Plots	Using the *INTERPOL = BOX* *<option(s)><00…25>* in the SYMBOL statement	Using VBOX, HBOX plot statement
Multiple Box Plots in One Page	Using PROC GREPLAY: must save each individual box plot in a category first.	Using PROC SGPANEL with PANEL by statement; no need to save individual box plot.
Pros	Easier to specify the options to produce whiskers with different length and outliers.	Can use the CATEGORY = option and GROUP = option to visualize the distribution of treatment groups with respect to baseline IOP categories.
Cons	No *CATEGORY = option* or *GROUP = option*, need to manipulate to create an index variable to visualize the distribution of treatment groups with respect to baseline IOP categories.	No options to specify the whisker length in SAS 9.3. This option is available in SAS 9.4.

9.6 References

Matange, S. 2013. "Custom Box Plots." SAS Institute Inc. blogs, March 24, http://blogs.sas.com/content/graphicallyspeaking/2013/03/24/custom-box-plots/.

SAS Institute Inc. 2012a. *SAS/GRAPH® 9.3: Reference,* 3rd ed. Cary, NC: SAS Institute Inc.

SAS Institute Inc. 2012b. *SAS® 9.3 ODS Graphics: Procedures Guide,* 3rd ed. Cary, NC: SAS Institute Inc.

Wicklin, R. 2012. "What Is the Difference between Categories and Groups in PROC SGPLOT?" SAS Institute Inc. blogs, August 22, http://blogs.sas.com/content/iml/2012/08/22/categories-vs-groups-in-proc-sgplot/.

9.7 Appendix: SAS Programs for Producing the Sample Figures

```
****************************************************************;
* Program Name: Chapter 9 Box Plots.sas                       *;
* Descriptions: Producing the following sample figures in     *;
* chapter 9                                                   *;
* - Figure 9.1 Treatment Effect by Baseline IOP at Week 12    *;
* - Figure 9.2 Treatment Effect by Baseline IOP and Visit     *;
****************************************************************;
options mprint symbolgen nodate nonumber validvarname = v7
orientation = landscape;
%let pgmname = Chapter 9 Box Plots.sas;
%let pgmloc = C:\SASBook\SAS Programs;
%let outloc = C:\SASBook\Sample Figures\Chapter 9;
%let pgmpth = &pgmloc.\&pgmname. &sysdate9. &systime. SAS
V&sysver.;

** Set-up macro variables for data simulation;
%LET SD = 3.5;
%let sitenum = 10;
%let seed = 08;
%let subjnum = 500;

proc format;
  value trtdf
    1, 4, 7 = 'Drug A'
    2, 5, 8 = 'Drug B'
    OTHER = ' ';
  value visdf
    1 = 'Baseline'
    2 = 'Week 4'
    3 = 'Week 8'
    4 = 'Week 12';
```

```
   value hrdf
      1 = 'Hour 0'
      2 = 'Hour 2'
      3 = 'Hour 8';
   value bsldf
      1 = 'BSL IOP: < 24'
      2 = 'BSL IOP: > = 24 to < = 26'
      3 = 'BSL IOP: > 26'
      OTHER = ' ';
run;

** Generate the required number of subjects and randomly
assign to 2 treatment groups;
data subj;
   do i = 1 to &subjnum.;
      subjid = 1000+ i;
      if ranuni (&seed.) < 0.5 then trtnum = 1;
      else trtnum = 2;
      output;
   end;
   drop i;
run;

** Set up the IOP Values based on the trt assignment and
visits/timepoints;
data iop;
   set subj;
   do i = 1 to 4; ** 4 visits;
      do j = 1 to 3; ** 3 timepoints/visit;
         visit = i;
         hour = j;
         if i = 1 then do; ** Baseline;
            if j = 1 then iop = round((RANNOR(&seed.)* &SD.
+ 25),.1); ** Hour 0;
            if j = 2 then iop = round((RANNOR(&seed.)* &SD.
+ 23),.1); ** Hour 2;
            if j = 3 then iop = round((RANNOR(&seed.)* &SD.
+ 22),.1); ** Hour 8;
         end;
         else if i > 1 and trtnum = 1 then do; ** Post-baseline:
drug A;
            if j = 1 then iop = round((RANNOR(&seed.)* &SD.
+ 17.5),.1); ** Hour 0;
            if j = 2 then iop = round((RANNOR(&seed.)* &SD.
+ 16.5),.1); ** Hour 2;
            if j = 3 then iop = round((RANNOR(&seed.)* &SD.
+ 16.2),.1); ** Hour 8;
         end;
```

```
        else if i > 1 and trtnum = 2 then do; ** Post-baseline:
drug B;
          if j = 1 then iop = round((RANNOR(&seed.)* &SD.
+ 20),.1); ** Hour 0;
            if j = 2 then iop = round((RANNOR(&seed.)* &SD.
+ 19),.1); ** Hour 2;
            if j = 3 then iop = round((RANNOR(&seed.)* &SD.
+ 18.7),.1); ** Hour 8;
      end;
      output;
    end;
  end;
  drop i j;
  format visit visdf. hour hrdf. trtnum trtdf.;
run;

proc sort data = iop;
  by subjid visit hour;
run;

** Mean diurnal IOP at each visit: mean of hours 0, 2 and 8
average eye IOP at baseline;
proc transpose data = iop out = iop_t;
  by subjid trtnum visit;
  var iop;
  id hour;
run;

data iop_MnDiur;
  set iop_t;
  iop_MnDiur = round (mean (Hour_0, Hour_2, Hour_8),.01);
  drop _NAME_ HOUR_:;
run;

** baseline mean diurnal IOP and category;
data mndiur_bsl;
  set iop_MnDiur;
  where visit = 1;
  if iop_MnDiur < 24 then bsl_c = 1;
  if 24 < = iop_MnDiur < = 26 then bsl_c = 2;
  if iop_MnDiur > 26 then bsl_c = 3;
  rename iop_MnDiur = mndiur_bsl;
  format bsl_c bsldf.;
  drop visit;
run;

** change from baseline;
data mndiur_chg;
  merge iop_MnDiur (where = (visit > 1)) mndiur_bsl;
  by subjid;
```

```
   chg = iop_MnDiur - mndiur_bsl;
run;

proc sort data = mndiur_chg;
   by visit bsl_c trtnum;
run;

data mndiur_chg2;
   set mndiur_chg;
   if bsl_c = 1 and trtnum = 1 then index = 1;
   if bsl_c = 1 and trtnum = 2 then index = 2;
   if bsl_c = 2 and trtnum = 1 then index = 4;
   if bsl_c = 2 and trtnum = 2 then index = 5;
   if bsl_c = 3 and trtnum = 1 then index = 7;
   if bsl_c = 3 and trtnum = 2 then index = 8;
run;

proc sort data = mndiur_chg2;
   by visit bsl_c trtnum;
run;

%LET FONTNAME = Times;
%LET DRIVER = PSCOLOR;%LEt EXT = PS;
goptions
   reset     = all
   GUNIT     = PCT
   rotate    = landscape
   gsfmode   = replace
   gsfname   = GSASFILE
   device    = &DRIVER
   lfactor   = 1
   hsize     = 8 in
   horigin   = 0 in
   vsize     = 6 in
   vorigin   = 6 in
   ftext     = "&FONTNAME"
   htext     = 10pt
   ftitle    = "&FONTNAME"
   htitle    = 10pt;

SYMBOL H = 2 CV = Black    CO = Black INTERPOL = BOXT10 W = 3
bwidth = 5 VALUE = CIRCLE;
axis1 major = none minor = none order = (0 to 8.5 by.5)
   label = (h = 2.5 font = "&FONTNAME" "Treatment")
   reflabel = (position = top c = blue font = "&FONTNAME" h = 2.5
j = r
   "BSL IOP: < 24 (mm Hg)" "BSL IOP: 24 to 26 (mm Hg)" "BSL
IOP: > 26 (mm Hg)");
```

```
axis2 order = (-15 to 5 by 5) minor = none label = (a = 90
h = 2.5
  font = "&fontname" "Change from Baseline at Week 4 (mm Hg)");
title1 "Treatment Effect by Baseline IOP at Week 4";
footnote1 "&pgmpth.";
FILENAME GSASFILE "&OUTLOC.\Week 4.&EXT.";
proc gplot data = mndiur_chg2 gout = work.bplot;
  where visit = 2;
  plot chg * index/vaxis = axis2 haxis = axis1 href = 0 href = 3
href = 6 name = "WK4";
  format index trtdf.;
run;

axis2 order = (-15 to 5 by 5) minor = none label = (a = 90
h = 2.5
  font = "&fontname" "Change from Baseline at Week 8 (mm Hg)");
title1 "Treatment Effect by Baseline IOP at Week 8";
FILENAME GSASFILE "&OUTLOC.\Week 8.&EXT.";
proc gplot data = mndiur_chg2 gout = work.bplot;
  where visit = 3;
  plot chg * index/vaxis = axis2 haxis = axis1 href = 0 href = 3
href = 6 name = "WK8";
  format index trtdf.;
run;

axis2 order = (-15 to 5 by 5) minor = none label = (a = 90
h = 2.5
  font = "&fontname" "Change from Baseline at Week 12 (mm Hg)");
title1 "Figure 9.1 Treatment Effect by Baseline IOP at Week 12";
FILENAME GSASFILE "&OUTLOC.\Week 12.&EXT.";
proc gplot data = mndiur_chg2 gout = work.bplot;
  where visit = 4;
  plot chg * index/vaxis = axis2 haxis = axis1 href = 0 href = 3
href = 6 name = "WK12";
  format index trtdf.;
run;
quit;

** Put the 3 figures in one page using PROC GREPLAY;
title1 "Figure 9.2 Treatment Effect by Baseline IOP at each
Visit";
footnote1 "&pgmpth.";
FILENAME GSASFILE "&OUTLOC.\Figure 9.2.&EXT.";
proc greplay igout = work.bplot;
  tc = sashelp.templt;
  template = v3;
  ** Replay into the chosen template;
  treplay 1:WK4 2:WK8 3:WK12;
run;
quit;
```

```
**************************************************************;
* Produce Box plots in SGPLOT and SGPANEL                 *;
**************************************************************;
%LET OUTPUTFMT = PS;
ods listing style = JOURNAL3 gpath = "&outloc.";
ods graphics/reset = all width = 8in height = 6in noborder
OUTPUTFMT = &OUTPUTFMT. imagename = "Fig_9_1";
title1 "Figure 9.1 Treatment Effect by Baseline IOP at Week 12";
footnote1 "&pgmpth.";
proc sgplot data = mndiur_chg2;
  where visit = 4;
  vbox chg/category = bsl_c group = trtnum;
  keylegend/position = bottom noborder;
  xaxis VALUES = (0 to 8.5 by.5) label = "Baseline IOP
Category";
  yaxis VALUES = (-15 to 5 by 5) label = "Change from Baseline
(mm Hg)";
  format bsl_c bsldf.;
run;
quit;

ods graphics/reset = all width = 8in height = 6in noborder
OUTPUTFMT = &OUTPUTFMT. imagename = "Fig_9_2";
title1 "Figure 9.2 Treatment Effect by Baseline IOP and
Visit";
title2 "Lattice by Visit";
footnote1 "&pgmpth.";
proc sgpanel data = mndiur_chg2;
  panelby visit/novarname layout = ROWLATTICE onepanel;
  vbox chg/category = bsl_c group = trtnum;
  keylegend/position = bottom noborder;
  colaxis VALUES = (0 to 8.5 by.5) label = "Baseline IOP
Category";
  rowaxis VALUES = (-15 to 5 by 5) label = "Change from
Baseline (mm Hg)";
  format bsl_c bsldf.;
run;
quit;

** The codes below produce box plots with whiskers displaying
10th and 90th percentiles;
** Summary for change from baseline;
proc means data = mndiur_chg noprint;
  class visit trtnum bsl_c;
  var chg;
  output out = chg_stat mean = Mean median = Median q1 = Q1
q3 = Q3 p10 = P10 p90 = P90;
run;
```

```
data chg_sum;
  set chg_stat;
  where visit > 1 and trtnum ne. and bsl_c ne.;
  format bsl_c bsldf. mean 4.1;
  meanlabelpos = mean+0.75;
  drop _FREQ_ _TYPE_;
run;

** Make custom box plots to extend the whiskers to the 10th
and 90th percentiles;
ods html close;
ods listing style = htmlblue gpath = "&outloc." image_dpi = 150;
ods graphics/reset = all width = 8in height = 6in noborder
imagename = "Fig_9_1b";
title1 "Figure 9.1 Treatment Effect by Baseline IOP at Week 12";
footnote1 "&pgmpth.";
proc sgplot data = chg_sum nocycleattrs;
  where visit = 4;
  highlow x = bsl_c high = p90 low = p10/group = trtnum
groupdisplay = cluster
    clusterwidth = 0.7;
  highlow x = bsl_c high = q3 low = median/group = trtnum type
= bar
    groupdisplay = cluster grouporder = ascending
clusterwidth = 0.7
    barwidth = 0.7 name = 'a';
  highlow x = bsl_c high = median low = q1/group = trtnum type
= bar
    groupdisplay = cluster grouporder = ascending
clusterwidth = 0.7 barwidth = 0.7;
  scatter x = bsl_c y = mean/group = trtnum groupdisplay =
cluster grouporder = ascending
    clusterwidth = 0.7 markerattrs = (size = 10);
  keylegend 'a';
  xaxis label = "Baseline IOP Category";
  yaxis grid label = "Change from Baseline (mm Hg)";
  format bsl_c bsldf.;
run;

ods graphics/reset = all width = 8in height = 6in noborder
imagename = "Fig_9_2b";
title1 "Figure 9.2 Treatment Effect by Baseline IOP and Visit";
title2 "Lattice by Visit";
footnote1 "&pgmpth.";
proc sgpanel data = chg_sum nocycleattrs;
  panelby visit/novarname layout = ROWLATTICE onepanel;
  highlow x = bsl_c high = p90 low = p10/group = trtnum
groupdisplay = cluster
    clusterwidth = 0.7;
```

```
  highlow x = bsl_c high = q3 low = median/group = trtnum type
= bar
    groupdisplay = cluster grouporder = ascending
clusterwidth = 0.7
    barwidth = 0.7 name = 'a';
  highlow x = bsl_c high = median low = q1/group = trtnum type
= bar
    groupdisplay = cluster grouporder = ascending
clusterwidth = 0.7 barwidth = 0.7;
  scatter x = bsl_c y = mean/group = trtnum groupdisplay =
cluster grouporder = ascending
    clusterwidth = 0.7 markerattrs = (size = 10);
  keylegend 'a';
  colaxis label = "Treatment by Baseline IOP Category";
  rowaxis grid label = "Change from Baseline (mm Hg)";
  format bsl_c bsldf.;
run;
quit;
```

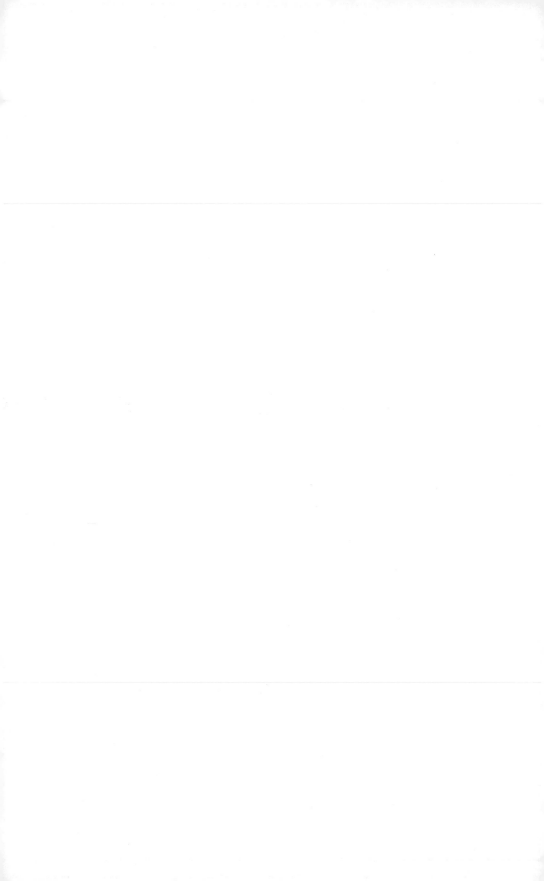

10

Forest Plots

10.1 Introduction

Forest plots show the estimates (e.g., mean values, odds ratios, hazard ratios, etc.) and the amount of variation (e.g., 95% confidence intervals) at different endpoints (measurement time, studies, adverse event, etc.). Forest plots in various forms have been published for more than 20 years, but have gained identity and popularity in the past 15 years (Bursac, 2010). In clinical research, forest plots can be used to display the mean and its associated 95% confidence interval (CI) for treatment difference from different endpoints in efficacy analyses and to display the odds ratio and its associated 95% CI for different adverse events (AE) in safety data analyses.

Forest plots differ from the thunderstorm scatter plots introduced in Chapter 6 in that forest plots display the estimates and the amount of variation, while thunderstorm scatter plots display repeated data with two or more values on the y-axis corresponding to one value on the x-axis. Forest plots are composed of scatters showing the estimates and lines extending in both directions from the estimates to the range of variation (e.g., 95% CI), and the plots can be displayed either horizontally or vertically. Thunderstorm scatter plots, on the other hand, are composed of scatters showing two or more repeated values of the same subject with lines connecting the values, which form raindrops, and are displayed vertically.

10.2 Application Examples

To illustrate the application and production of forest plots, two examples are presented in this chapter based on clinical research in the glaucoma therapeutic area. Figure 10.1, used in efficacy analysis, shows the treatment differences and their 95% CI at all post-baseline visits and hours after treatment. Figure 10.2, used in safety analysis, displays the relative risks (RR) and their associated 95% CI for some major AEs as well as overall in a study.

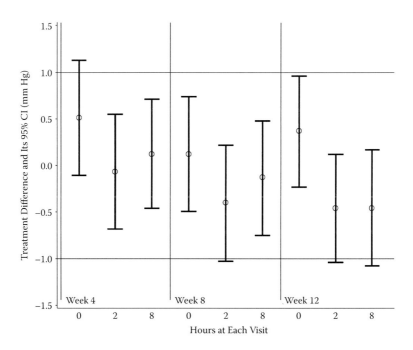

FIGURE 10.1
Treatment difference and its 95% CI.

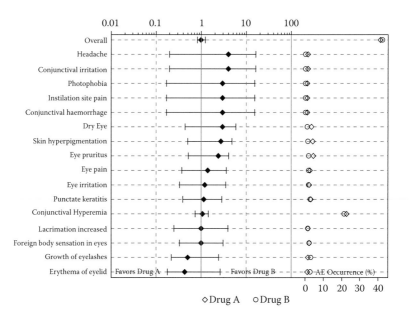

FIGURE 10.2
Relative risk and its 95% CI with AE occurrence (%).

10.2.1 Example 1: Mean Estimates and 95% CI for Treatment Difference at Each Post-baseline Visit and Hour

Let's design a virtual clinical trial to demonstrate that Drug A is equivalent to Drug B in IOP reduction in patients with glaucoma or ocular hypertension. Let's assume an equivalence margin of ± 1.0 mm Hg, and a standard deviation (SD) of 3.5 to calculate the sample size. To demonstrate that Drug A is equivalent to Drug B in IOP reduction, 259 patients per treatment group and 518 overall in the study are required. The sample size calculations considered a one-sided $\alpha = 0.025$, 80% power, and no expected difference between treatment groups, and were determined using the commercial software PASS (Hintze, 2006).

The least squares (LS) mean and its 95% CI are estimated based on an analysis of covariance (ANCOVA) model with the treatment and investigation site as the fixed effect and the baseline IOP as the covariate. The forest plot with the LS mean and its 95% CI at each post-baseline visit and hour (hours 0, 2, and 8 of weeks 4, 8, and 12) are displayed in Figure 10.1. The forest plot allows us to visualize the treatment difference estimates at each post-baseline time point and easily count the time points with the 95% CI falling within the +/−1.0 mm Hg equivalence margin and helps us to decide whether Drug A is equivalent to Drug B.

10.2.2 Example 2: Relative Risk and 95% CI for Different Adverse Events

Both Figures 10.2 and 10.3 display AEs with relative risk and 95% CI on the left side as a forest plot for the most frequently occurred AEs and overall. Figure 10.2 displays the actual AE occurrence rate (%), while Figure 10.3 displays the actual RR values and the lower and the upper bounds of the 95% CI in the right side panel. Both variations of the forest plots are commonly used in clinical study safety data analyses.

10.3 Producing the Sample Figures

10.3.1 Data Structure and SAS Annotated Dataset

Table 10.1 includes the estimated treatment difference and its 95% CI derived from the ANCOVA model based on the simulated IOP data for Drugs A and B. The dataset is used to produce the forest plot in Figure 10.1. Variables *Index, Index_l,* and *Index-r* are added to position the time points on the x-axis to display the 95% CI bars.

Table 10.2 has the AE occurrence rate data (*PCT_A* and *PCT_B*), RR, and 95% CI (*LowerCL* and *UpperCL*) and is used to produce Figures 10.2 and 10.3.

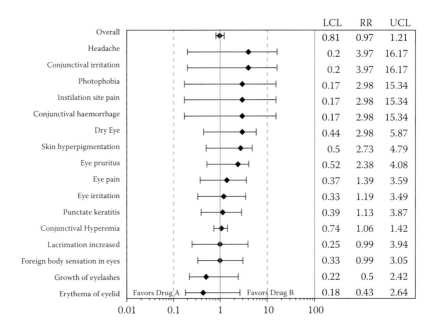

FIGURE 10.3

Relative risk and its 95% CI with actual values.

TABLE 10.1

Estimated Treatment Difference and 95% CI

visit	hour	diff	lcl	ucl	index	index_l	index_r
Week 4	Hour 0	0.51	−0.11	1.13	1	0.85	1.15
Week 4	Hour 2	−0.07	−0.68	0.55	2	1.85	2.15
Week 4	Hour 8	0.12	−0.46	0.71	3	2.85	3.15
Week 8	Hour 0	0.12	−0.49	0.74	4	3.85	4.15
Week 8	Hour 2	−0.4	−1.03	0.22	5	4.85	5.15
Week 8	Hour 8	−0.13	−0.75	0.48	6	5.85	6.15
Week 12	Hour 0	0.37	−0.23	0.96	7	6.85	7.15
Week 12	Hour 2	−0.46	−1.04	0.12	8	7.85	8.15
Week 12	Hour 8	−0.46	−1.08	0.17	9	8.85	9.15

10.3.2 Notes to SAS Programs

The SAS programs that were used to produce the three sample forest plots using both the GPLOT and SGPLOT procedures are included in the Appendix (Section 10.6). The program consists of two parts, with the first part producing the forest plot for treatment differences and its 95% CI for IOP reduction (Figure 10.1), and the second part producing the forest plots for RR and its 95% CI for AEs (Figures 10.2 and 10.3).

TABLE 10.2

Relative Risk and 95% Confidence Interval for AEs

AE	PCT_A	PCT_B	RRisk	LowerCL	UpperCL	AEId	RR	LCL	UCL
Erythema of eyelid	1.15	2.71	0.43	0.18	2.64	1	RR	LCL	UCL
Growth of eyelashes	1.54	3.1	0.5	0.22	2.42	2	RR	LCL	UCL
Foreign body sensation in eyes	2.31	2.33	0.99	0.33	3.05	3	RR	LCL	UCL
Lacrimation increased	1.54	1.55	0.99	0.25	3.94	4	RR	LCL	UCL
Conjunctival Hyperemia	22.69	21.32	1.06	0.74	1.42	5	RR	LCL	UCL
Punctate keratitis	3.08	2.71	1.13	0.39	2.87	6	RR	LCL	UCL
Eye irritation	2.31	1.94	1.19	0.33	3.49	7	RR	LCL	UCL
Eye pain	2.69	1.94	1.39	0.37	3.59	8	RR	LCL	UCL
Eye pruritus	4.62	1.94	2.38	0.52	4.08	9	RR	LCL	UCL
Skin hyperpigmentation	4.23	1.55	2.73	0.5	4.79	10	RR	LCL	UCL
Dry Eye	3.46	1.16	2.98	0.44	5.87	11	RR	LCL	UCL
Conjunctival hemorrhage	1.15	0.39	2.98	0.17	15.34	12	RR	LCL	UCL
Instillation site pain	1.15	0.39	2.98	0.17	15.34	13	RR	LCL	UCL
Photophobia	1.15	0.39	2.98	0.17	15.34	14	RR	LCL	UCL
Conjunctival irritation	1.54	0.39	3.97	0.2	16.17	15	RR	LCL	UCL
Headache	1.54	0.39	3.97	0.2	16.17	16	RR	LCL	UCL
Overall	41.15	42.25	0.97	0.81	1.21	17	RR	LCL	UCL

10.3.2.1 First Part Main Sections and Features

10.3.2.1.1 Dataset Simulation

An IOP dataset is simulated to include 518 subjects, with each subject randomly assigned to one of the 8 sites and to either Drug A or Drug B at a 1:1 ratio at each site. IOP values are assigned based on baseline and post-baseline visit and hours, with no expected difference between the two treatment groups at all visits and hours.

10.3.2.1.2 Data Analyses and Manipulation

- The IOP change from baseline is calculated for post-baseline visits and hours.
- The estimated LS mean and its 95% CI are derived from the ANCOVA model with treatment and investigator site as fixed effects and baseline IOP as the covariate. PROC GLM is used and the LS Mean and its 95% CI are output in a dataset *LSMeanDiffCL*.

```
ods listing close;
ods trace on;
ods output LSMEanDiffCL = LSMEanDiffCL;
proc glm data = iop_chg;
   by visit hour;
   class site trtnum;
   model iop_chg = site trtnum iop_bsl/ss3;
   lsmeans trtnum/cl pdiff;
run;
quit;
ods output close;
ods trace off;
ods listing;
```

- ODS LISTING CLOSE is used to turn off any listing destinations for any files produced by PROC GLM; ODS TRACE ON is used to display all the output datasets produced in PROC GLM in the LOG window; ODS OUTPUT LSMEanDiffCL = LSMEanDiffCL is used to save the output dataset LSMEanDiffCL in the WORK directory.
- The SAS Annotate dataset is created to draw the lines to form the forest plots: one line connecting the lower and upper 95% CI, and two lines to draw short bars at the lower and upper 95% CI points.

10.3.2.1.3 Producing Figure 10.1 in GPLOT

The GPLOT procedure in SAS/GRAPH is used to produce Figure 10.1. The scatters (points) for the LS mean are produced using the *PLOT diff * index* statement with the I = None option in the symbol statement; the 95% CI bars and the lines connecting the two bars are drawn using the annotated dataset described previously.

```
SYMBOL H = 4 C = BLACK CO = BLACK I = NONE font = 'albany
amt/unicode' VALUE = '25cb'x;

PROC GPLOT DATA = diff_cl;
   PLOT diff * index/HAXIS = AXIS2 VAXIS = AXIS1 href = 0.5
href = 3.5 href = 6.5 vref = -1 vref = 1
     noframe ANNOTATE = ANNOT;
   FORMAT index indexdf.;
RUN;
QUIT;
```

10.3.2.1.4 Producing Figures 10.1 in SGPLOT

Two versions of Figure 10.1 are produced using the SGPLOT procedure. One is produced using the *SCATTER* plot statement to display the LS mean value and using the three *HighLow* plot statements to draw lines to connect the 95% CI points (available in SAS 9.3). The other is produced using only one *SCATTER* plot statement with *YERRORLOWER* and *YERRORUPPER* options to display the 95% CI lines (available in SAS 9.2).

```
** Option 1: Using Scatter and Highlow statements;
proc sgplot data = diff_cl noautolegend nocycleattrs;
   ** Display the mean value in a scatter plot;
   scatter x = index y = diff/markerattrs = (symbol =
circle size = 10);

   ** Display the 95% CI using HighLow plot;
   highlow x = index low = lcl high = ucl/type = line
lineattrs = (thickness = 2);

   ** Display the short lines at the interval points using
HighLow plot;
   highlow y = lcl low = index_l high = index_r/type = line;
   highlow y = ucl low = index_l high = index_r/type = line;

   * Draw the reference lines and other details;
   refline 0.5 3.5 6.5/axis = x lineattrs = (thickness = 2);
   refline -1 1/axis = y lineattrs = (thickness = 2);
   inset 'Week 4' /position = bottomleft;
   inset 'Week 8' /position = bottom;
   inset 'Week 12'/position = bottomright;
```

```
   xaxis VALUES = (0.5 to 9.5 by.5) label = "Hours at Each
Visit" display = (NOTICKS);
   yaxis VALUES = (-1.5 to 1.5 by.5) label = "Treatment
Difference and Its 95% CI (mm Hg)";
   FORMAT index indexdf.;
run;

** Option 2: Using one Scatter statement with yerrorlower
and yerrorupper options;
proc sgplot data = diff_cl noautolegend;
   ** Display the mean value using scatter plot;
   scatter x = index y = diff/markerattrs = (symbol =
circle size = 10) yerrorlower = lcl yerrorupper = ucl;

   * Draw the reference lines and other details;
   refline 0.5 3.5 6.5/axis = x lineattrs = (thickness = 2);
   refline -1 1/axis = y lineattrs = (thickness = 2);
   inset 'Week 4' /position = bottomleft;
   inset 'Week 8' /position = bottom;
   inset 'Week 12'/position = bottomright;

   xaxis VALUES = (0.5 to 9.5 by.5) label = "Hours at Each
Visit" display = (NOTICKS);

   yaxis VALUES = (-1.5 to 1.5 by.5) label = "Treatment
Difference and Its 95% CI (mm Hg)";
   FORMAT index indexdf.;
run;
```

10.3.2.2 Second Part Main Sections and Features

- Enter the data RR to include the AE occurrence rate for each drug, the RR, and its 95% CI (the upper and lower levels). Please note that the data are used for demonstration purposes and are not real data.

- The SGPLOT procedure is used to produce the forest plots of RR and its 95% CI for AEs. The GPLOT is not used here because it requires much more care and effort.

- Figure 10.2 is produced to include RR and its 95% CI on the left side of the graph and the actual AE occurrence rate (%) on the right side of the graph. The RR and its 95% CI, and the AE occurrence rates are positioned in two x-axes.

```
proc sgplot data = rr noautolegend nocycleattrs;
   * Draw reference lines for each AE;
   refline AE/lineattrs = (thickness = 1 pattern =
shortdash);
```

```
   * Display AE occurrence (%) using scatter plot;
   scatter y = AE x = PCT_A/markerattrs = (symbol =
diamond color = blue size = 10) name = "DRG1" legendlabel
= "DRUG A";
   scatter y = AE x = PCT_B/markerattrs = (symbol = circle
color = red size = 10) name = "DRG2" legendlabel = "DRUG B";

   * Display the AE RR and its 95%;
   scatter y = AE x = rrisk/markerattrs = (symbol =
diamondfilled size = 10) xerrorlower = lowercl xerrorupper
= uppercl x2axis ;

   * Draw reference lines and insets;
   refline 1 100/axis = x2;
   refline 0.01 0.1 10/axis = x2 lineattrs = (pattern =
shortdash);
   refline 0.01 0.1 10/axis = x2 lineattrs = (pattern =
shortdash);

   inset 'Favors Drug A' /position = bottomleft;
   inset 'Favors Drug B' /position = bottom;
   inset 'AE Occurrence (%)' /position = bottomright;

   xaxis offsetmin = 0.7 display = (nolabel);
   x2axis type = log offsetmin = 0 offsetmax = 0.35
min = 0.01 max = 100 minor display = (nolabel);
   yaxis display = (noticks nolabel);
   keylegend "DRG1" "DRG2"/noborder;
run;
```

- Figure 10.3 is produced to include RR and its 95% CI on the left
  side of the graph and the actual values for the RR and its 95% CI on
  the right side of the graph. The RR and its 95% CI, and the actual
  values are positioned in two x-axes. The reference lines for AEs,
  as shown in Figure 10.2, are not used in Figure 10.3 to avoid inter-
  fering with readability of the numbers for RR and 95% CI in the
  right side panel.

```
proc sgplot data = rr noautolegend nocycleattrs;
   * Display the AE RR and its 95% CI;
   scatter y = AE x = rrisk/markerattrs = (symbol =
diamondfilled size = 10) xerrorlower = lowercl xerrorupper
= uppercl;

   * Display the actual values for RR and its 95% CI on
X2 axis;
```

```
scatter y = AE x = lcl/markerchar = lowercl x2axis;
scatter y = AE x = rr/markerchar = rrisk x2axis;
scatter y = AE x = ucl/markerchar = uppercl x2axis;

* Draw reference lines and insets;
refline 1 100/axis = x;
refline 0.01 0.1 10/axis = x lineattrs = (pattern =
shortdash);

inset 'Favors Drug A' /position = bottomleft;
inset 'Favors Drug B' /position = bottom;

xaxis type = log offsetmin = 0 offsetmax = 0.35
min = 0.01 max = 100 minor display = (nolabel);
x2axis offsetmin = 0.7 display = (noticks nolabel);
yaxis display = (noticks nolabel);
run;
quit;
```

- The TRANSPANCY feature of reference lines are not used because this feature will result in bitmap image format figures in SAS 9.3, even though the vector format output (PS, EMF, etc.) are used and saved. The codes above are modified based on the SAS programs to produce forest plots in Mantange and Heath's book on statistical graphics procedures (Mantange and Heath, 2011).

10.4 Summary and Discussion

Forest plots can be produced using both GPLOT and SGPLOT procedures. However, the more sophisticated forest plots, like Figures 10.2 and 10.3, can be much more easily produced using SGPLOT with less coding and effort. Table 10.3 summarizes the main features in producing the forest plots using the two procedures.

10.5 References

Bursac, Z. 2010. "Creating Forest Plots from Pre-computed Data Using PROC SGPLOT and Graph Template Language." In *SAS Global Forum 2010 Proceedings*, Cary, NC: SAS Institute Inc., http://support.sas.com/resources/papers/proceedings10/195-2010.pdf.

TABLE 10.3

Comparing PROC GPLOT and PROC SGPLOT in Producing Forest Plots

Features	GPLOT in SAS/GRAPH	SGPLOT in ODS Graphics
Scatters (for mean values)	PLOT Y*X statement with INTERPOL = NONE in the SYMBOL statement.	SCATTER plot statement
Variation range, e.g., 95% CI	Using an annotated dataset	Using HighLow statements or the yerrorlower/yerrorupper options in a Scatter plot statement
Inset	Not available	A good feature to add inserts within a figure; only applied to the SGPLOT procedure.
Pros	Easy to use and produce a simple forest plot together with an annotated dataset.	Easier to produce more complicated forest plots like Figures 10.2 and 10.3, which include a forest plot and additional data (statistics, actual AE rate, etc.).
Cons	More difficult to produce complicated forest plots.	Might need a learning curve to be comfortable with it.

Hintze, J. 2006. NCSS, PASS, and GESS [computer program]. Kaysville, UT: NCSS.
Mantange, S., and Heath, D. 2011. *Statistical Graphics Procedures by Example: Effective Graphs Using SAS®*. Cary, NC: SAS Institute Inc.

10.6 Appendix: SAS Programs for Producing the Sample Figures

```
***************************************************************;
* Program Name: Chapter 10 Forest Plots.sas                   *;
* Descriptions: Producing the following sample figures in     *;
* Chapter 10                                                  *;
* - Figure 10.1 Mean and 95% CI for treatment difference      *;
* by visit and hour                                           *;
* - Figure 10.2 AEs: OR and 95% CI                            *;
* - Figure 10.3 Relative Risk and 95% CI with Actual Values   *;
* Shown                                                       *;
***************************************************************;
options mprint symbolgen nodate nonumber validvarname = v7
orientation = landscape;
%let pgmname = Forest Plots.sas;
%let pgmloc = C:\SASBook\SAS Programs;
%let pgmpth = &pgmloc.\&pgmname. &sysdate9. &systime. SAS
V&sysver.;
%let outloc = C:\SASBook\Sample Figures\Chapter 10;
```

```
** Set-up the site, subject number and SD for data simulation;
%let sitenum = 10;
%let subjnum = 518;
%let SD = 3.5;
%let seed = 100;

proc format;
  value trtdf
    1 = 'New Drug'
    2 = 'Active Control'
    OTHER = ' ';
  value visdf
    1 = 'Baseline'
    2 = 'Week 4'
    3 = 'Week 8'
    4 = 'Week 12';
  value hrdf
    1 = 'Hour 0'
    2 = 'Hour 2'
    3 = 'Hour 8';
  value indexdf
    1, 4, 7 = '0'
    2, 5, 8 = '2'
    3, 6, 9 = '8'
    OTHER = ' ';
run;

** Generate the required number of subjects;
data subj;
  do i = 1 to &subjnum.;
    subjid = 1000+ i;
    shuffle = ranuni (&seed.);
    output;
  end;
  drop i;
run;

proc sort data = subj;
  by shuffle;
run;

** Randomly assign the subjects to each site;
** Within each site randomly assign 2 treatment groups;
data site_subj;
  set subj;
  if shuffle <.1 then do;
    site = 1;
    if ranuni (&seed.) < 0.5 then trtnum = 1;
  else trtnum = 2;
  end;
```

```
  else if.1 < = shuffle <.30 then do;
    site = 2;
    if ranuni (&seed.) < 0.5 then trtnum = 1;
    else trtnum = 2;
  end;
  else if.30 < = shuffle <.35 then do;
    site = 3;
    if ranuni (&seed.) < 0.5 then trtnum = 1;
    else trtnum = 2;
  end;
  else if.35 < = shuffle <.5 then do;
    site = 4;
    if ranuni (&seed.) < 0.5 then trtnum = 1;
    else trtnum = 2;
  end;
  else if.5 < = shuffle <.6 then do;
    site = 5;
    if ranuni (&seed.) < 0.5 then trtnum = 1;
    else trtnum = 2;
  end;
  else if.6 < = shuffle <.8 then do;
    site = 6;
    if ranuni (&seed.) < 0.5 then trtnum = 1;
    else trtnum = 2;
  end;
  else if.8 < = shuffle <.85 then do;
    site = 7;
    if ranuni (&seed.) < 0.5 then trtnum = 1;
    else trtnum = 2;
  end;
  else if.85 < = shuffle < = 1.0 then do;
    site = 8;
    if ranuni (&seed.) < 0.5 then trtnum = 1;
    else trtnum = 2;
  end;
run;

proc freq data = site_subj noprint;
  table trtnum/out = subj_trt;
run;

** Save the subject number at each treatment group to macro
variables;
data _null_;
  set subj_trt;
  if trtnum = 1 then call symput ("N_Trt1", put(count, 3.0));
  if trtnum = 2 then call symput ("N_Trt2", put(count, 3.0));
run;
```

```
** Set up the IOP Values based on the trt assignment and
visits/timepoints;
data iop;
  set site_subj;
  do i = 1 to 4; ** 4 visits;
    do j = 1 to 3; ** 3 timepoints/visit;
      visit = i;
      hour = j;
      if i = 1 then do; ** Baseline;
        if j = 1 then iop = round((RANNOR(&SEED.)* &SD.
+ 25),.1); ** Hour 0;
        if j = 2 then iop = round((RANNOR(&SEED.)* &SD.
+ 23),.1); ** Hour 2;
        if j = 3 then iop = round((RANNOR(&SEED.)* &SD.
+ 22),.1); ** Hour 8;
      end;
      else if i > 1 then do; ** Post-baseline;
        if j = 1 then iop = round((RANNOR(&SEED.)* &SD.
+ 17.5),.1); ** Hour 0;
        if j = 2 then iop = round((RANNOR(&SEED.)* &SD.
+ 16.5),.1); ** Hour 2;
        if j = 3 then iop = round((RANNOR(&SEED.)* &SD.
+ 16.2),.1); ** Hour 8;
      end;
      output;
    end;
  end;
  drop i j shuffle;
  format visit visdf. hour hrdf. trtnum trtdf.;
run;

proc sort data = iop;
  by site subjid visit hour;
run;

** Baseline IOP;
proc sort data = iop out = iop_bsl (drop = visit rename = (iop
= iop_bsl));
  by subjid hour;
  where visit = 1;
run;

proc sort data = iop out = iop_pbsl;
  by subjid hour;
  where visit > 1;
run;

data iop_chg;
  merge iop_pbsl iop_bsl;
  by subjid hour;
```

```
   iop_chg = iop - iop_bsl;
run;

proc sort data = iop_chg;
   by visit hour;
run;

ods listing close;
ods trace on;
ods output LSMEanDiffCL = LSMEanDiffCL;
proc glm data = iop_chg;
   by visit hour;
   class site trtnum;
   model iop_chg = site trtnum iop_bsl/ss3;
   lsmeans trtnum/cl pdiff;
run;
quit;

ods output close;
ods trace off;
ods listing;

data diff_cl;
   set LSMEanDiffCL;
   diff = Round(Difference,.01);
   lcl = round(LowerCL,.01);
   ucl = round(UpperCL,.01);
   if visit = 2 and hour = 1 then index = 1;
   if visit = 2 and hour = 2 then index = 2;
   if visit = 2 and hour = 3 then index = 3;
   if visit = 3 and hour = 1 then index = 4;
   if visit = 3 and hour = 2 then index = 5;
   if visit = 3 and hour = 3 then index = 6;
   if visit = 4 and hour = 1 then index = 7;
   if visit = 4 and hour = 2 then index = 8;
   if visit = 4 and hour = 3 then index = 9;
   index_l = index -.15;
   index_r = index +.15;
   keep visit hour diff lcl ucl index:;
run;

* Make the SAS Annotate data set macros available for use;
%ANNOMAC;

* Generate an annotation data set that draw indicators for
means and their 95% CI;
DATA ANNOT;
   SET diff_cl;
   %DCLANNO;
   RETAIN SEMULT 1 NUM_OFFSET.1 SHIFT_VAL.20;
```

```
   SIZE = 2; HSYS = '4'; XSYS = '2'; YSYS = '2';
   * draw the vertical line connecting lcl and ucl *;
   %LINE(index, lcl, index, ucl, black, 1, SIZE);
   * lower bar for 95% CI;
   %LINE(index-SHIFT_VAL, lcl, index+SHIFT_VAL, lcl, black, 1,
SIZE);
   %* upper bar for 95% CI;
   %LINE(index-SHIFT_VAL, ucl, index+SHIFT_VAL, ucl, black, 1,
SIZE);
RUN;

%LET FONTNAME = Times;
%LET DRIVER = PSCOLOR;%LEt EXT = PS;
goptions
   reset    = all
   GUNIT    = PCT
   rotate   = landscape
   gsfmode  = replace
   gsfname  = GSASFILE
   device   = &DRIVER
   lfactor  = 1
   hsize    = 8 in
   horigin  = 0.5 in
   vsize    = 6.5 in
   vorigin  = 0.75 in
   ftext    = "&FONTNAME"
   htext    = 10pt
   ftitle   = "&FONTNAME"
   htitle   = 10pt
;

SYMBOL H = 4 C = BLACK CO = BLACK I = NONE font = 'albany amt/
unicode' VALUE = '25cb'x;
AXIS1 OFFSET = (1,1) ORDER = (-1.5 TO 1.5 BY.5) LABEL = (FONT
= "&FONTNAME." h = 2.5
   ANGLE = 90 "Treatment Difference and Its 95% CI (mm Hg)")
VALUE = (H = 2.5)MINOR = NONE;
AXIS2 OFFSET = (1,1) MAJOR = NONE MINOR = NONE ORDER = (.5 TO
9.5 BY.5)
   LABEL = (FONT = "&FONTNAME." h = 2.5 'Hours at Each Visit')
   reflabel = (position = bottom c = green h = 2.5 j = r FONT =
"&FONTNAME." "Week 4" "Week 8" "Week 12");
FILENAME GSASFILE "&outloc./Figure 10.1.&EXT.";

title1 "Figure 10.1 Treatment Difference and 95% CI (mm Hg)";
footnote1 "Note: Treatment Difference and its 95% CI at each
hour (represented by the dot and bars) are based on";
footnote2 " the ANCOVA model with treatment and investigator
as fixed effects and baseline IOP as the covariate.";
footnote3 "&pgmpth.";
```

```
PROC GPLOT DATA = diff_cl;
   PLOT diff * index/HAXIS = AXIS2 VAXIS = AXIS1 href = 0.5
href = 3.5 href = 6.5 vref = -1 vref = 1 noframe
   ANNOTATE = ANNOT;
   FORMAT index indexdf.;
RUN;
QUIT;

***************************************************************;
* Produce the Forest Plots in SGPLOT and SGPANEL          *;
***************************************************************;
%LET OUTPUTFMT = PS;
ods listing gpath = "&outloc.";
ods graphics/reset = all width = 8in height = 6in noborder
OUTPUTFMT = &OUTPUTFMT. imagename = "FigSG_10_1";
title1 "Figure 10.1 Treatment Difference and 95% CI (mm Hg)";
footnote1 "Note: Treatment Difference and its 95% CI at each
hour (represented by the dot and bars) are based on";
footnote2 " the ANCOVA model with treatment and investigator
as fixed effects and baseline IOP as the covariate.";
footnote3 "&pgmpth.";

** Option 1: Using Scatter and Highlow statements;
proc sgplot data = diff_cl noautolegend nocycleattrs;
   ** Display the mean value in a scatter plot;
   scatter x = index y = diff/markerattrs = (symbol = circle
size = 10);

   ** Display the 95% CI using HighLow plot;
   highlow x = index low = lcl high = ucl/type = line lineattrs
= (thickness = 2);

   ** Display the short lines at the interval points using
HighLow plot;
   highlow y = lcl low = index_l high = index_r/type = line;
   highlow y = ucl low = index_l high = index_r/type = line;

   * Draw the reference lines and other details;
   refline 0.5 3.5 6.5/axis = x lineattrs = (thickness = 2);
   refline -1 1/axis = y lineattrs = (thickness = 2);
   inset 'Week 4' /position = bottomleft;
   inset 'Week 8' /position = bottom;
   inset 'Week 12'/position = bottomright;

   xaxis VALUES = (0.5 to 9.5 by.5) label = "Hours at Each Visit"
display = (NOTICKS);
   yaxis VALUES = (-1.5 to 1.5 by.5)
     label = "Treatment Difference and Its 95% CI (mm Hg)";
   FORMAT index indexdf.;
run;
```

```
** Option 2: Using one Scatter statement with yerrorlower and
yerrorupper options;
proc sgplot data = diff_cl noautolegend;
  ** Display the mean value using scatter plot;
  scatter x = index y = diff/markerattrs = (symbol = circle
size = 10) yerrorlower = lcl yerrorupper = ucl;

  * Draw the reference lines and other details;
  refline 0.5 3.5 6.5/axis = x lineattrs = (thickness = 2);
  refline -1 1/axis = y lineattrs = (thickness = 2);
  inset 'Week 4' /position = bottomleft;
  inset 'Week 8' /position = bottom;
  inset 'Week 12'/position = bottomright;

  xaxis VALUES = (0.5 to 9.5 by.5) label = "Hours at Each Visit"
display = (NOTICKS);
  yaxis VALUES = (-1.5 to 1.5 by.5) label = "Treatment
Difference and Its 95% CI (mm Hg)";
  FORMAT index indexdf.;
run;

** Data with mean and 95% CI for treatment difference;
data rr;
  input AE $1-30 PCT_A PCT_B RRisk LowerCL UpperCL;
  AEId = _N_;

  * Set up columns to create the stat tables;
  RR = 'RR'; LCL = 'LCL'; UCL = 'UCL';
  datalines;
Erythema of eyelid               1.15   2.71  0.43  0.18   2.64
Growth of eyelashes              1.54   3.10  0.50  0.22   2.42
Foreign body sensation in eyes   2.31   2.33  0.99  0.33   3.05
Lacrimation increased            1.54   1.55  0.99  0.25   3.94
Conjunctival Hyperemia          22.69  21.32  1.06  0.74   1.42
Punctate keratitis               3.08   2.71  1.13  0.39   2.87
Eye irritation                   2.31   1.94  1.19  0.33   3.49
Eye pain                         2.69   1.94  1.39  0.37   3.59
Eye pruritus                     4.62   1.94  2.38  0.52   4.08
Skin hyperpigmentation           4.23   1.55  2.73  0.50   4.79
Dry Eye                          3.46   1.16  2.98  0.44   5.87
Conjunctival haemorrhage         1.15   0.39  2.98  0.17  15.34
Instilation site pain            1.15   0.39  2.98  0.17  15.34
Photophobia                      1.15   0.39  2.98  0.17  15.34
Conjunctival irritation          1.54   0.39  3.97  0.20  16.17
Headache                         1.54   0.39  3.97  0.20  16.17
Overall                         41.15  42.25  0.97  0.81   1.21
;
run;
```

```
ods listing style = JOURNAL3 gpath = "&outloc.";
ods graphics/reset = all width = 8in height = 6in noborder
OUTPUTFMT = &OUTPUTFMT. imagename = "Fig10_2";
title1 "Figure 10.2 Relative Risk and 95% CI with AE
Occurrence Rate";
footnote1 "&pgmpth.";
proc sgplot data = rr noautolegend nocycleattrs;
   * Draw reference lines for each AE;
   refline AE/lineattrs = (thickness = 1 pattern = shortdash);

   * Display AE occurrence (%) using scatter plot;
   scatter y = AE x = PCT_A/markerattrs = (symbol = diamond color
= blue size = 10)
      name = "DRG1" legendlabel = "DRUG A";
   scatter y = AE x = PCT_B/markerattrs = (symbol = circle color
= red size = 10)
      name = "DRG2" legendlabel = "DRUG B";

   * Display the AE RR and its 95%;
   scatter y = AE x = rrisk/markerattrs = (symbol = diamondfilled
size = 10) xerrorlower = lowercl xerrorupper = uppercl x2axis ;

   * Draw reference lines and insets;
   refline 1 100/axis = x2;
   refline 0.01 0.1 10/axis = x2 lineattrs = (pattern =
shortdash);
   refline 0.01 0.1 10/axis = x2 lineattrs = (pattern =
shortdash);

   inset 'Favors Drug A' /position = bottomleft;
   inset 'Favors Drug B' /position = bottom;
   inset 'AE Occurrence (%)' /position = bottomright;

   xaxis offsetmin = 0.7 display = (nolabel);
   x2axis type = log offsetmin = 0 offsetmax = 0.35 min = 0.01
max = 100 minor display = (nolabel);
   yaxis display = (noticks nolabel);
   keylegend "DRG1" "DRG2"/noborder;
run;

ods listing style = JOURNAL3 gpath = "&outloc.";
ods graphics/reset = all width = 8in height = 6in noborder
OUTPUTFMT = &OUTPUTFMT. imagename = "Fig10_3";
title1 "Figure 10.3 Relative Risk and 95% CI with Actual Values
Shown";
footnote1 "&pgmpth.";
proc sgplot data = rr noautolegend nocycleattrs;
```

```
* Display the AE RR and its 95% CI;
  scatter y = AE x = rrisk/markerattrs = (symbol =
diamondfilled size = 10) xerrorlower = lowercl xerrorupper =
uppercl;

  * Display the actual values for RR and its 95% CI on X2 axis;
  scatter y = AE x = lcl/markerchar = lowercl x2axis;
  scatter y = AE x = rr/markerchar = rrisk x2axis;
  scatter y = AE x = ucl/markerchar = uppercl x2axis;

  * Draw reference lines and insets;
  refline 1 100/axis = x;
  refline 0.01 0.1 10/axis = x lineattrs = (pattern =
shortdash);
  inset 'Favors Drug A'/position = bottomleft;
  inset 'Favors Drug B'/position = bottom;
  xaxis type = log offsetmin = 0 offsetmax = 0.35 min = 0.01
max = 100 minor display = (nolabel);
  x2axis offsetmin = 0.7 display = (noticks nolabel);
  yaxis display = (noticks nolabel);
run;
quit;
```

11

Survival Plots

11.1 Introduction

Survival data refer to observations of subjects after intervention and over a certain length of time (even lifetime) until the occurrence of an event. In clinical research, an event can be the time from treatment to death in cancer research, the time from treatment in a bone fracture in osteoporosis clinical trials, and so on. PROC LIFETEST, one of the survival analysis procedures in SAS/STAT, computes the Kaplan–Meier estimates of the survival functions and compares survival plots (curves) between different treatment groups (Kuhfeld and So, 2013). The log-rank test is usually used to analyze the intervention effects in different groups.

Different variations of the survival plots can be produced directly using the LIFETEST procedure. A more customized version can be created using the GPLOT or SGPLOT procedures with the output dataset generated by running the LIFETEST procedure first.

This chapter illustrates how to produce survival plots using SGPLOT as well as directly from the LIFETEST procedure.

11.2 Application Examples

To illustrate the application and production of survival plots, we generate breast cancer survival data for 100 subjects. Let's assume that after a subject is diagnosed with breast cancer, the subject is treated with either Drug A or Drug B and followed up for a period of 10 years. The survival times (1 to 10 years) are recorded with a status of either 1 or 0, with 1 indicating death and 0 for a censored time (lost to follow-up or still alive at the end of the 10-year period). Figure 11.1 displays the survival rates while Figure 11.2 displays the failure rates for subjects in the two treatment groups; both figures were produced directly using the LIFETEST procedure. Figure 11.3 is produced in SGPLOT using the output dataset produced in LIFETEST and is

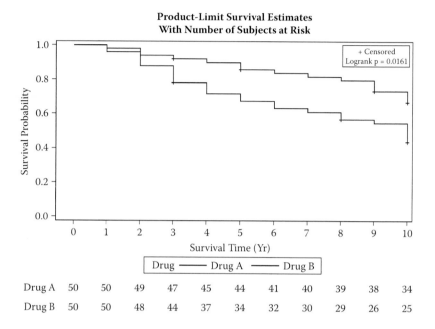

FIGURE 11.1
A survival plot produced using the LIFETEST procedure directly.

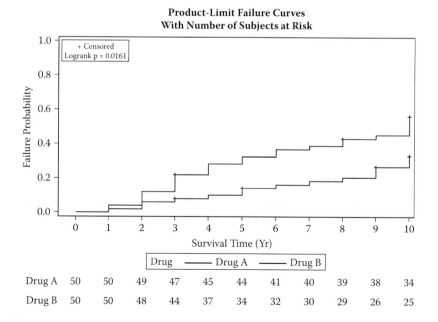

FIGURE 11.2
A failure plot produced using the LIFETEST procedure directly.

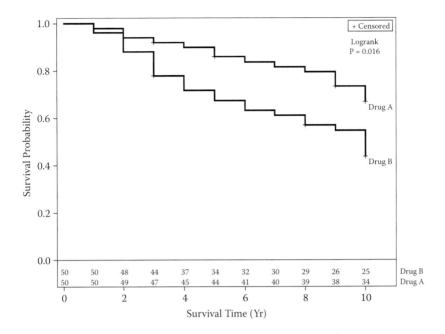

FIGURE 11.3
A survival plot produced using the SGPLOT procedure.

a more customized version of Figure 11.1. The figures demonstrate that the cancer subjects in the Drug A group have better survival rates than those in the Drug B group, especially after two years of treatment.

11.3 Producing the Sample Figures

11.3.1 Data Structure

Table 11.1 includes survival data for 10 subjects including subject ID, treatment group, survival time, and status. The SurvivalPlot dataset produced by the LIFETEST procedure is used to produce Figure 11.3 using the SGPLOT procedure. Part of the dataset with the survival probability, number of subjects at risk, event, and censored from year 0 to 10 is shown in Table 11.2.

11.3.2 Notes to SAS Programs

The SAS program that produced the three sample survival (or failure) plots using the LIFETEST and SGPLOT procedures is provided in the Appendix (Section 11.6). The program consists of two parts: the first part uses the LIFETEST procedure to produce the survival and the failure plots directly

TABLE 11.1

Breat Cancer Survival Data for 10 Subjects

SubjID	Drug	Survival Time (year)	Status: 1 = Death, 0 = Censored
1	1	10	0
2	1	9	1
3	1	5	1
4	1	9	1
5	1	10	1
6	1	3	1
7	1	10	0
8	1	7	1
9	1	10	0
10	1	10	0

with different options; the second part uses the SGPLOT to produce a more customized survival plot.

11.3.2.1 Producing Survival Plots Using the LIFETEST Procedure Directly

The LIFETEST procedures can be used directly to produce the survival (or failure) plots with different variations (Kuhfeld and So, 2013). Figure 11.1 is a survival plot with the at-risk number of subjects for each year displayed outside the body of the curve itself. It is done by using the following codes.

```
** Produce the survival plot and output the survival data
from the LIFETEST procedure;
ods graphics on;
ods listing gpath = "&outloc.";
ods output Survivalplot = Survivalplot HomTests =
HomTests;
ods select survivalplot(persist) failureplot(persist);
ods graphics/reset = all width = 8in height = 6in
noborder OUTPUTFMT = EMF imagename = "FigSG_11_1";
title1 "Figure 11.1. Survival Plots Produced in
LIFETEST";
title2 "- with the at-risk table displayed outside the
body of the graph";
footnote "&pgmpth.";
proc lifetest data = BCANCER plots = survival(test
atrisk(atrisktick outside(0.15)) = 0 to 10 by 1);
   time Year * Status(0);
   strata Drug;
run;
```

TABLE 11.2

Part of Survival Plot Data Produced by LIFETEST (0 to 5 years)

Time	Survival	AtRisk	Event	Censored	tAtRisk	Stratum	StratumNum
0	—	50	—	—	0	Drug A	1
0	—	50	—	—	0	Drug B	2
0	1	50	0	—	—	Drug A	1
0	1	50	0	—	—	Drug B	2
1	—	50	—	—	1	Drug A	1
1	—	50	—	—	1	Drug B	2
1	0.96	50	2	—	—	Drug B	2
1	0.98	50	1	—	—	Drug A	1
2	—	48	—	—	2	Drug B	2
2	—	49	—	—	2	Drug A	1
2	0.88	48	4	—	—	Drug B	2
2	0.94	49	2	—	—	Drug A	1
3	—	44	—	—	3	Drug B	2
3	—	44	0	0.78	—	Drug B	2
3	—	47	—	—	3	Drug A	1
3	—	47	0	0.92	—	Drug A	1
3	0.78	44	5	—	—	Drug B	2
3	0.92	47	1	—	—	Drug A	1
4	—	37	—	—	4	Drug B	2
4	—	45	—	—	4	Drug A	1
4	0.716756757	37	3	—	—	Drug B	2
4	0.899555556	45	1	—	—	Drug A	1
5	—	34	—	—	5	Drug B	2
5	—	44	—	—	5	Drug A	1
5	—	44	0	0.858666667	—	Drug A	1
5	0.674594595	34	2	—	—	Drug B	2
5	0.858666667	44	2	—	—	Drug A	1

- The ODS GRAPHICS ON statement is used to turn on the ODS Graphics.
- The ODS OUTPUT is used to save the two output datasets (SurvivalPlot and HomTests) in the Work directory.
- The SurvivalPlot dataset contains variables for survival probability, number of subjects at risk, event, and censored from year 0 to 10 (Table 11.2), and the HomTests dataset contains information for homogeneity test between the two drugs, with test statistics (Chi-square) and p-values from Log-Rank, Wilcoxon, and −2Log(LR) tests.
- The ODS SELECT statement is used to display only the survival and failure time plots. If not used, the survival estimate tables will be produced by default together with the plots.

- ODS GRAPHICS is used to specify the plot dimensions, output format, and filename, and save the figure in the designated location.
- The number of subjects at risk for each year and the location of the display is controlled by the options for plots: plots = survival(test atrisk(atrisktick outside (0.15)) = 0 to 10 by 1). It is located 15% outside the survival curves at each year from years 0 to 10.

Figure 11.2 is a failure plot, displaying the number of subjects failed (death in this example) by time. It is produced using the codes below with the "failure" option in LIFETEST.

```
** Produce the failure plot from the LIFETEST procedure;
ods graphics/reset = all width = 8in height = 6in
noborder OUTPUTFMT = EMF imagename = "FigSG_11_2";
title1 "Figure 11.2. Failure Plots Produced in LIFETEST";
title2 "- with the at-risk table displayed outside the
body of the graph";
footnote "&pgmpth.";
proc lifetest data = BCANCER plots = survival(failure
test atrisk(atrisktick outside(0.15)) = 0 to 10 by 1);;
  time Year * Status(0);
  strata Drug;
run;
```

11.3.2.2 Producing Survival Plots Using the SGPLOT Procedure with the Output Dataset from LIFETEST

We can produce more customized survival plots using GPLOT or SGPLOT with the Survivalplot dataset generated by the LIFETEST procedure. The figure produced can also be saved in different listing format files (PS, EMF, etc.). Figure 11.3 is a survival plot with the number of subjects survived (at risk) for each year displayed in a different axis than the survival probability. The codes are based on a sample in the book *Statistical Graphics Procedures by Example* (Matange and Heath, 2011).

```
ods select all;
ods graphics/reset = all width = 8in height = 6in
noborder OUTPUTFMT = EMF imagename = "FigSG_11_3";
title1 "Figure 11.3. Survival Plots Produced in Proc
SGPLOT";
title2 "- with the at-risk table displayed at the bottom
of the grap in diffrenet axis";
footnote "&pgmpth.";
```

```
proc sgplot data = Survivalplot2 sganno = anno_label;
   * Use the step plot with curvelabels for label purpose;
   step x = time y = survival/group = stratum curvelabel
      lineattrs = (thickness = 2) name = 'survival';
   * Draw censored observations;
   scatter x = time y = censored/markerattrs = (symbol =
plus size = 10) name = 'censored';
   scatter x = time y = censored/markerattrs = (symbol =
plus size = 10) GROUP = stratum;
   * Draw the At Risk values in the 2nd y-axis;
   scatter x = tatrisk y = stratumnum/markerchar = atrisk
y2axis group = stratumnum;
   * Reference line and legend;
   refline 0;
   keylegend 'censored'/location = inside position =
topright;
   * Set axis properties and offsets to separate the curve
from the table with number at risk;
   yaxis offsetmin = 0.1 min = 0;
   y2axis offsetmax =.94 display = (nolabel noticks)
valueattrs = (size = 8);
   format stratumNum drugdf.;
run;
```

- The ODS SELECT ALL statement is used to suppress the effect of *"ods select survivalplot(persist) failureplot(persist)"* to allow the figure produced in SGPLOT to display in the output window.

- The SGANNO facility is used to display the Log-rank test p-values in the plot.

- The survival plot is a step plot. The STEP plot statement in the SGPLOT procedure is used to display the step functions at each year by treatment group.

- Two SCATTER plot statements are used to indicate the censored time points by group and with a name "censored" for legend purposes.

- Another SCATTER plot statement is used to display the at-risk table in the second y-axis by drug (stratum).

- YAXIS and Y2AXIS statements are used to specify the options for the two y-axes, especially for the offset locations. You might need to experiment to choose the offset levels to display the plot and the table in the desired locations.

The ODS RTF is used to save the three plots in an RTF file with the options *nogtitle nogfootnote*, which put the titles and footnotes in the headers and footers instead of the body of the figure image. The figures can be copied and pasted to other Microsoft applications, like Word, PowerPoint, and so on. Similarly, the figures can be saved in the PDF file using ODS PDF statement.

The survival plots can also be produced using the GPLOT procedure, but more codes and effort are needed, including use of the annotate facility and Proc Greplay to overlay the survival curves and the at-risk table. The codes are not provided here.

11.4 Summary and Discussion

Survival plots can be produced using the LIFETEST procedure directly with the GPLOT or SGPLOT procedures. The STEP and SSCATTER plot statements in SGPLOT can be used to produce more customized survival plots and allow the figure to be saved in different listing graphic outputs (PS, EMF, etc.). Survival plots produced in SGPLOT can have better quality than those directly produced in LIFETEST.

11.5 References

Kuhfeld, W., and So, Y. 2013. "Creating and Customizing the Kaplan–Meier Survival Plot in PROC LIFETEST." SAS Global Forum 2013 Proceedings, Cary, NC: SAS Institute Inc., http://support.sas.com/resources/papers/proceedings13/427-2013.pdf.

Matange, S., and Heath, D. 2011. *Statistical Graphics Procedures by Example: Effective Graphs Using SAS®*. Cary, NC: SAS Institute Inc.

11.6 Appendix: SAS Programs for Producing the Sample Figures

```
*************************************************************;
* Program Name: Chapter 11 Survival Plots.sas              *;
* Descriptions: Producing the following sample figures in  *;
* chapter 11                                               *;
* - Figure 11.1. Survival Plots Produced in LIFETEST       *;
* - Figure 11.2. Failure Plots Produced in LIFETEST        *;
* - Figure 11.3. Survival Plots Produced in Proc SGPLOT    *;
*************************************************************;
options mprint symbolgen nodate nonumber validvarname = v7
orientation = landscape;
```

```
%let pgmname = Chapter 11 Survival plot for breast cancer.sas;
%let pgmloc = C:\SASBook\SAS Programs;
%let pgmpth = &pgmloc.\&pgmname. &sysdate9. &systime. SAS
V&sysver.;
%let outloc = C:\SASBook\Sample Figures\Chapter 11;

proc format;
  value drugdf
    1 = 'Drug A'
    2 = 'Drug B';
run;

data BCANCER;
  input Subjid Drug Year Status @@;
  format Drug drugdf.;
  label Year = 'Survival Time (Yr)'
    status = 'Status: 1 = death, 0 = cencored';
  datalines;
1 1 10 0 2 1 9 1 3 1 5 1 4 1 9 1 5 1 10 1
6 1 3 1 7 1 10 0 8 1 7 1 9 1 10 0 10 1 10 0
11 1 10 0 12 1 3 0 13 1 6 1 14 1 10 0 15 1 10 0
16 1 1 1 17 1 10 0 18 1 10 0 19 1 10 0 20 1 5 0
21 1 10 0 22 1 10 0 23 1 10 0 24 1 10 0 25 1 4 1
26 1 10 0 27 1 10 1 28 1 10 0 29 1 10 0 30 1 10 0
31 1 2 1 32 1 10 0 33 1 10 0 34 1 10 0 35 1 8 1
36 1 10 0 37 1 10 0 38 1 9 0 39 1 10 0 40 1 10 0
41 1 10 0 42 1 2 1 43 1 10 0 44 1 10 1 45 1 10 0
46 1 10 0 47 1 5 1 48 1 10 0 49 1 10 0 50 1 9 1
51 2 4 1 52 2 10 1 53 2 2 1 54 2 5 1 55 2 10 0
56 2 1 1 57 2 10 0 58 2 9 1 59 2 3 1 60 2 10 0
61 2 3 0 62 2 6 1 63 2 10 0 64 2 3 1 65 2 10 0
66 2 4 1 67 2 10 0 68 2 2 1 69 2 10 0 70 2 5 1
71 2 10 1 72 2 8 0 73 2 10 0 74 2 2 1 75 2 10 0
76 2 6 1 77 2 10 0 78 2 3 0 79 2 10 1 80 2 8 1
81 2 10 0 82 2 3 1 83 2 10 0 84 2 7 1 85 2 10 1
86 2 10 0 87 2 3 1 88 2 10 0 89 2 10 0 90 2 2 1
91 2 10 0 92 2 4 1 93 2 10 0 94 2 10 1 95 2 8 1
96 2 10 0 97 2 1 1 98 2 10 0 99 2 10 0 100 2 3 1
;
run;

ods rtf file = "&outloc.\Chapter 11 Survival Plots.rtf" style
= minimal nogtitle nogfootnote;

** Produce the survival plot and output the survival data from
the LIFETEST procedure;
ods graphics on;
ods listing gpath = "&outloc.";
ods output Survivalplot = Survivalplot HomTests = HomTests;
ods select survivalplot(persist) failureplot(persist);
```

```
ods graphics/reset = all width = 8in height = 6in noborder
OUTPUTFMT = EMF imagename = "FigSG_11_1";
title1 "Figure 11.1. Survival Plots Produced in LIFETEST";
title2 "- with the at-risk table displayed outside the body
of the graph";
footnote "&pgmpth.";
proc lifetest data = BCANCER plots = survival(test
atrisk(atrisktick outside(0.15)) = 0 to 10 by 1);
  time Year * Status(0);
  strata Drug;
run;

** Produce the failure plot from the LIFETEST procedure;
ods graphics/reset = all width = 8in height = 6in noborder
OUTPUTFMT = EMF imagename = "FigSG_11_2";
title1 "Figure 11.2. Failure Plots Produced in LIFETEST";
title2 "- with the at-risk table displayed outside the body
of the graph";
footnote "&pgmpth.";
proc lifetest data = BCANCER plots = survival(failure test
atrisk(atrisktick outside(0.15)) = 0 to 10 by 1);;
time Year * Status(0);
strata Drug;
run;

****************************************************************;
** Produce the survival plot using SGPLOT                    *;
****************************************************************;
* Save the p-value from the Logrank test to a macro variable;
data _NULL_;
  set homtests;
  if test = "Log-Rank" then call symput ('LR_PVal',
put(ProbChiSq, 5.3));
run;

* get rid of duplicates in the SurvivalPlot dataset;
proc sort data = survivalplot out = survivalplot2 nodup;
  by Time Survival AtRisk Event Censored tAtRisk Stratum
StratumNum;
run;

** Annotated dataset to displying the Logrank P value;
data anno_label;
  function = "text"; x1 = 10; y1 =.9; x1space = "datavalue";
y1space = "datavalue";
  label = "Logrank P = &LR_PVal."; textcolor = "black"; output;
run;
```

```
ods select all;
ods graphics/reset = all width = 8in height = 6in noborder
OUTPUTFMT = EMF imagename = "FigSG_11_3";
title1 "Figure 11.3. Survival Plots Produced in Proc SGPLOT";
title2 "- with the at-risk table displayed at the bottom
of the grap in diffrenet axis";
footnote "&pgmpth.";
proc sgplot data = Survivalplot2 sganno = anno_label;
  * Use the step plot with curvelabels for label purpose;
  step x = time y = survival/group = stratum curvelabel
    lineattrs = (thickness = 2) name = 'survival';
  * Draw censored observations;
  scatter x = time y = censored/markerattrs = (symbol = plus
size = 10) name = 'censored';
  scatter x = time y = censored/markerattrs = (symbol = plus
size = 10) GROUP = stratum;
  * Draw the At Risk values in the 2nd y-axis;
  scatter x = tatrisk y = stratumnum/markerchar = atrisk
y2axis group = stratumnum;
  * Reference line and legend;
  refline 0;
  keylegend 'censored'/location = inside position = topright;
  * Set axis properties and offsets to separate the curve from
the table with number at risk;
  yaxis offsetmin = 0.1 min = 0;
  y2axis offsetmax =.94 display = (nolabel noticks) valueattrs
= (size = 8);
  format stratumNum drugdf.;
run;
quit;
ods rtf close;
```

12

Waterfall Plots and Histograms

12.1 Introduction

This chapter discusses and illustrates how to produce two types of plots that are not as commonly used as those described in the previous chapters—waterfall plots and histograms. A *waterfall plot* is an ordered chart where each subject is symbolized by a vertical bar, representing the change with respect to a reference value (Nieto and Gómez, 2010).

Histograms provide data visualization of continuous numerical data by indicating the number of data points that lie within a range of values, called a *class* or a *bin*. The frequency of the data that fall in each class is depicted by the use of a rectangular bar. The rectangles of a histogram are drawn so that they touch each other to indicate that the original variable is continuous.

Histograms are different from bar charts, although both use rectangular bars to represent data. In a histogram, it is the area of the bar that represents the value, not the height as in a bar chart (Wikipedia). Bar charts are used to display data at the nominal level of measurement, that is, categorical data, and the categories are the classes for a bar chart. In a bar chart, the bars are commonly rearranged in order of decreasing height. Histograms, on the other hand, are used to display data that is continuous, or at the ordinal level of measurement, and the classes are ranges of values. The bars in a histogram cannot be rearranged and must be displayed in the order that the classes occur.

12.2 Application Examples

To illustrate the application and production of waterfall and histogram plots, two examples are presented in this chapter, one waterfall plot for intraocular pressure (IOP) change from baseline of 50 subjects and the other a histogram for age distribution by treatment and gender.

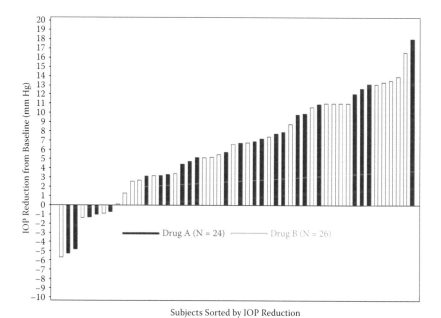

FIGURE 12.1
Waterfall plot for IOP reduction from baseline.

12.2.1 Waterfall Plot: Change from Baseline in IOP for 50 Subjects

Let's assume a Phase I clinical trial designed to compare the IOP reduction of drugs A and B in 50 subjects. To employ the principle of letting the data to speak for themselves, especially for a study with a relatively small number of subjects, a visualization plot to display all subjects IOP reduction from baseline at the endpoint visit would be a useful tool. Figure 12.1, a waterfall plot for the IOP reduction from baseline at week 12 displays the range of the IOP reduction and the relative position for all subjects in the two treatments. From the figure, we can see that there are 8 subjects with increased IOP after treatment (negative reduction: 5 from Drug A and 3 from Drug B) and overall, Drug B is more efficacious than Drug A.

12.2.2 Histogram Plot

Let's assume a clinical study with 200 subjects enrolled with ages between 18 and 80. We would like to see the distribution of subjects' age overall and by treatment group and gender. Histograms serve this purpose well. Figure 12.2 is a histogram showing the overall age distribution together with the density function, and Figure 12.3 is a multipanel plot for age distribution by gender and treatment group.

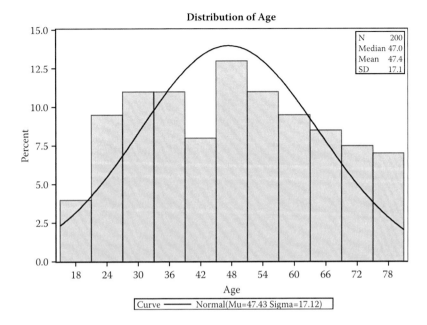

FIGURE 12.2
Overall age distribution for all subjects.

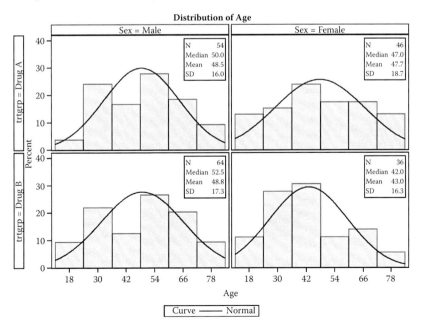

FIGURE 12.3
Age distribution by gender and treatment group.

12.3 Producing the Sample Figures

12.3.1 Data Structure and SAS Annotated Dataset

Table 12.1 displays the IOP reduction data structure for 10 subjects. Figure 12.1 was produced using the IOP reduction data for 50 subjects. The data are sorted by the IOP reduction value and have a variable "Subj_odr" containing the order to position the subject data on the x-axis.

Table 12.2 displays the age data structure for 10 subjects. Figures 12.2 and 12.3 were produced using the age data for a total of 200.

12.3.2 Notes to SAS Programs

The two SAS programs that were used to produce the sample plots using PROC GPLOT, PROC SGPLOT, and PROC UNIVARIATE are provided in the Appendix (Section 12.6).

TABLE 12.1

Part of the Subjects' IOP Reduction Data for Figure 12.1

subjid	trtgrp	iop_bsl	iop_w12	reduction	reduction1	reduction2	subj_odr
1020	Drug B	16.8	22.5	−5.7	—	−5.7	1
1050	Drug A	17.7	23	−5.3	−5.3	—	2
1014	Drug A	20.6	25.4	−4.8	−4.8	—	3
1008	Drug B	24.3	25.7	−1.4	—	−1.4	4
1040	Drug A	16.3	17.6	−1.3	−1.3	—	5
1039	Drug A	18.3	19.3	−1	−1	—	6
1017	Drug B	23.6	24.5	−0.9	—	−0.9	7
1001	Drug A	27.1	27.8	−0.7	−0.7	—	8
1026	Drug B	18.2	18.1	0.1	—	0.1	9
1009	Drug B	21.7	20.4	1.3	—	1.3	10

TABLE 12.2

Part of the Subjects' Age Data
for Figures 12.2 and 12.3

subjid	age	trtgrp	sex
1001	67	Drug A	Female
1002	57	Drug A	Male
1003	55	Drug B	Male
1004	36	Drug B	Female
1005	48	Drug A	Female
1006	35	Drug B	Male
1007	34	Drug B	Female
1008	46	Drug B	Female
1009	56	Drug B	Male
1010	50	Drug A	Female

12.3.2.1 Main Sections of the First Program

12.3.2.1.1 Dataset Simulation and Manipulation

- Fifty subjects' baseline and week 12 IOP data are simulated from a normal distribution with the preset mean and standard deviation (SD). Subjects are randomly assigned to either the Drug A or Drug B group at a 1:1 ratio.

- The number of subjects in each treatment group is calculated using PROC FREQ and saved in macro variables (N1 and N2) using the function CALL SYMPUT ().

- The IOP reduction from baseline at week 12 is calculated for all subjects and the data is sorted by the IOP reduction. Separate variables (*reduction1* and *reduction2*) are created for the IOP reduction for Drug A and Drug B. A variable *Subj_Odr* is generated to contain the relative position for IOP reduction from the smallest to the biggest on the x-axis in the waterfall plot.

- An SAS annotate dataset is created to produce the bars and symbols in a waterfall plot.

  - SAS provided macros *%ANNOMAC* and *%DCLANNO* are invoked first, so the *%bar()* macro can be called.

  - The *%bar()* macro is used to draw a rectangle or bar using two sets of x/y coordinates, which specify diagonal corners (SAS Institute, 2012). This macro has the syntax of %bar (*x1, y1, x2, y2, color, line, style*) with *x1* and *y1* used to specify the coordinates for the first corner of the bar, and *x2* and *y2* used to specify the coordinates for the second corner of the bar, which is diagonal to the first corner.

  - Besides producing the bars in the waterfall plot, the annotation data is also used to position the labels for the two treatment groups.

12.3.2.1.2 Producing the Waterfall Plot in GPLOT

PROC GPLOT is used produce the waterfall plot in Figure 12.1. The annotate dataset (anno) is used to generate the bars to create the waterfalls and labels for the treatment groups and the *Value = None* option is specified in the SYMBOL statement.

```
SYMBOL1 VALUE = None;
axis1 major = none minor = none order = (0 to 51 by 1)
   label = (h = 2.5 font = "&FONTNAME" "Subjects Sorted by
IOP Reduction");
axis2 order = (-10 to 20 by 1) minor = none label = (a = 90
r = 0 h = 2.5
```

```
    font = "&fontname" "IOP Reduction from Baseline (mm Hg)");

 title1 "Figure 12.1 Waterfall Plot: IOP Reduction from
 Baseline";
 footnote1 "&pgmpth.";
 FILENAME GSASFILE "&OUTLOC.\Figure 12.1.&EXT.";
 proc gplot data = iop_reduction;
   plot reduction*subj_odr/anno = anno haxis = axis1 vaxis
 = axis2 vref = 0;
   format subj_odr subjdf.;
 run;
 quit;
```

12.3.2.1.3 Reproduce the Waterfall Plot Using SGPLOT

A NEEDLE plot statement in SGPLOT can be used to easily produce the bars in a waterfall plot. A needle becomes a bar when the thickness of the line is increased (thickness = 6 is used in the example).

Needle plots use vertical line segments to connect each data point to a baseline (SAS Institute, 2012). It has the syntax of NEEDLE X = variable Y = numeric-variable </option(s)>;. In the following example, PATTERN = SOLID or DOT is used to specify the line pattern to distinguish the two drugs. A line pattern can be specified using either a word or a number, and the complete list of available patterns can be found in the *SAS ODS Graphics Procedures Guide* (SAS Institute, 2012).

```
 ods listing gpath = "&outloc.";
 ods graphics/reset = all width = 8in height = 6in
 noborder OUTPUTFMT = emf imagename = "FigSG_13_1";
 title1 &GRAPHTTL "Figure 12.1 Waterfall Plot: IOP
 Reduction from Baseline";
 footnote1 "&pgmpth.";
 proc sgplot data = iop_reduction;
   needle x = subj_odr y = reduction1/lineattrs =
 (thickness = 6 PATTERN = solid COLOR = red);
   needle x = subj_odr y = reduction2/lineattrs =
 (thickness = 6 PATTERN = DOT COLOR = blue);
   format subj_odr subjdf.;
   xaxis display = (nolabel);
   yaxis label = 'IOP Reduction from Baseline (mm Hg)'
 VALUES = (-10 to 20 by 1);
   label reduction1 = "Drug A (N = &N1.)" reduction2 =
 "Drug B (N = &N2.)";
   keylegend/location = inside position = bottom;
 run;
```

12.3.2.2 Main Sections of the Second Program

12.3.2.2.1 Dataset Simulation and Manipulation

- The age data for 200 subjects are simulated to include ages between 18 and 80 years. Subjects are assigned to two treatment groups as well as male and female.
- The subject number, the mean, median, and the standard deviation of age are calculated using Proc Means and saved into macro variables.

12.3.2.2.2 Producing the Histogram Using Proc Univariate

- PROC UNIVRIATE is used to produce two histograms, one for age distribution overall and the other for age distribution by treatment group and gender. An INSET statement is used to inset a table of summary of statistics in the graph at the specified position, including the sample size, median, mean, and the standard deviation.

```
title1 "Figure 12.2 Histogram: Distribution of Age
Overall";
footnote1 "&pgmpth.";
FILENAME GSASFILE "&OUTLOC.\Figure 12.2.&EXT.";
PROC UNIVARIATE DATA = age noprint;
   HISTOGRAM age/NORMAL CFILL = ltgray;
   INSET N = 'N' MEDIAN (5.1) MEAN (5.1) STD = "SD" (5.1)/
POSITION = ne;
RUN;

title1 "Figure 12.3 Histogram: Distribution of Age by
Treatment Group and Gender";
footnote1 "&pgmpth.";
FILENAME GSASFILE "&OUTLOC.\Figure 12.3.&EXT.";
PROC UNIVARIATE DATA = age noprint;
   class trtgrp sex;
   HISTOGRAM age/NORMAL CFILL = ltgray;
   INSET N = 'N' MEDIAN (5.1) MEAN (5.1) STD = "SD" (5.1)/
POSITION = ne;
RUN;
```

12.3.2.2.3 Producing the Histogram Using SGPLOT and SGPANEL

- A HISTOGRAM plot statement is used in PROC SGPLOT to reproduce the histogram for the overall age distribution, and a DENSITY statement is used to produce the density curve.

```
proc sgplot data = age noautolegend;
   histogram age/transparency =.5;
```

```
      density age;
      xaxis display = (nolabel);
      inset "N:&Num., Median:&MD., MEAN:&MN.,  SD:&SD."/
position = TOPRIGHT;
run;
```

- A HISTOGRAM plot statement together with PANELBY is used in PROC SGPANEL to reproduce the histogram for the age distribution by treatment group and gender. A DENSITY statement is used to produce the density curve.

```
proc sgpanel data = age noautolegend;
panelby sex trtgrp/layout = lattice novarname;
histogram age/transparency =.5;
density age;
rowaxis; colaxis;
run;
```

12.4 Summary and Discussion

Waterfall plots can be produced in both GPLOT and SGPLOT. The vertical bars in a waterfall plot can be produced using the %bar() function in the annotate facility in GPLOT and using the NEEDLE plot statement in SGPLOT. Using the NEEDLE statement in SGPLOT is an easier way to produce waterfall plots.

A histogram can be produced using PROC UNIVARIATE, PROC SGPLOT, and PROC SGPANEL. However, the insets of the summary statistics can be more easily done in the UNIVARIATE procedure.

12.5 References

Nieto, A., and Gómez, J. 2010. "Waterfall Plots: A Beautiful and Easy Way of Showing a Whole Picture of an Interesting Outcome." PhUSE 2010, Paper SP08, http://www.phusewiki.org/docs/2010/2010%20PAPERS/SP08%20Paper.pdf.

SAS Institute Inc. 2012. *SAS/GRAPH® 9.3: Reference,* 3rd ed. Cary, NC: SAS Institute Inc.

SAS Institute Inc. 2012. *SAS® 9.3 ODS Graphics: Procedures Guide,* 3rd ed. Cary, NC: SAS Institute Inc.

Taylor, Courtney. "What Is a Histogram?" About.com, About Education, http://statistics.about.com/od/HelpandTutorials/a/What-Is-A-Histogram.htm.

12.6 Appendix: SAS Programs for Producing the Sample Figures

12.6.1 Waterfall Plots

```
**************************************************************;
* Program Name: Chapter 13 Other Types of Plots:          *;
* Waterfall Plots.sas                                      *;
* Descriptions: Producing the following sample figures in  *;
* chapter 1                                                *;
* - Figure 12.1 Waterfall Plot: IOP Reduction from Baseline *;
**************************************************************;
options mprint symbolgen nodate nonumber validvarname = v7
orientation = landscape;
%let pgmname = Chapter 12 Other Types of Plots - Waterfall
Plots.sas;
%let pgmloc = C:\SASBook\SAS Programs;
%let outloc = C:\SASBook\Sample Figures\Chapter 12;
%let pgmpth = &pgmloc.\&pgmname. &sysdate9. &systime. SAS
V&sysver.;

** Set-up the site, subject number and SD for data simulation;
%let subjnum = 50;
%LET SD = 4;
%LET SEED = 12;

proc format;
  value trtdf
    1 = 'Drug A'
    2 = 'Drug B'
    OTHER = ' ';
  value sexdf
    1 = 'Male'
    2 = 'Female'
    OTHER = ' ';
  value subjdf
    0 - 100 = ' '
    OTHER = ' ';
run;

** Generate the required number of subjects;
data IOP;
  do i = 1 to &subjnum.;
    subjid = 1000+ i;
    shuffle = ranuni (&SEED.);
      if 0 < = shuffle < = 0.5 then do;
        trtgrp = 1;
        iop_bsl = round((RANNOR(&SEED.)* &SD. + 26),.1);
** BSL IOP for Drug A;
```

```
           iop_w12 = round((RANNOR(&SEED.)* &SD. + 20),.1);
** Wk12 IOP for Drug B;
       end;
       else if 0.5 < shuffle < = 1 then do;
          trtgrp = 2;
          iop_bsl = round((RANNOR(&SEED.)* &SD. + 26),.1);
** BSL IOP for Drug B;
          iop_w12 = round((RANNOR(&SEED.)* &SD. + 18),.1);
** WK12 IOP for Drug B;
       end;
       output;
  end;
  drop i;
  format trtgrp trtdf.;
run;

** Subject number by treatment;
proc freq data = IOP noprint;
  table trtgrp/out = subj_trt;
run;

** Save the subject number at each treatment group to macro
variables for later use;
data _null_;
  set subj_trt;
  if trtgrp = 1 then call symput ("N1", put(count, 3.0));
  if trtgrp = 2 then call symput ("N2", put(count, 3.0));
run;

data iop_reduction;
  set iop;
  reduction = iop_bsl-iop_w12; ** Reduction from baseline;
run;

proc sort data = iop_reduction;
  by reduction;
run;

data iop_reduction;
  set iop_reduction;
  if trtgrp = 1 then reduction1 = reduction; else reduction1
=.;
  if trtgrp = 2 then reduction2 = reduction; else reduction2
=.;
  subj_odr = _N_;
run;

* Make the SAS Annotate data set macros available for use;
%ANNOMAC;

data ANNO_bar;
  set iop_reduction;
```

```
   %DCLANNO;
   RETAIN SDMULT 1 NUM_OFFSET.1 SHIFT_VAL.15;
   SIZE = 3;
   HSYS = '4';
   XSYS = '2';
   YSYS = '2';
   if trtgrp = 1 then do;
      %bar(subj_odr, 0, subj_odr+.5, reduction, red, 1, 'S');
   end;
   if trtgrp = 2 then do;
      %bar(subj_odr, 0, subj_odr+.5, reduction, blue, 1, 'E');
   end;
   KEEP X Y FUNCTION COLOR LINE SIZE HSYS XSYS YSYS STYLE trtgrp;
RUN;

data anno_label;
   SIZE = 1.5;
   HSYS = '4';
   XSYS = '2';
   YSYS = '2';
   %bar(10, -3.25, 15, -3, red, 1, 'S');
   x = 20; y = -3; function = 'label'; color = 'red';
   text = "Drug A (N = &N1.)"; output;
   %bar(25, -3.25, 30, -3, blue, 1, 'E');
   x = 35; y = -3; function = 'label'; color = 'blue';
   text = "Drug B (N = &N2.)"; output;
run;

data anno;
   set ANNO_bar anno_label;
run;

%LET FONTNAME = Times;
%LET DRIVER = PSCOLOR;%LEt EXT = PS;
goptions
   reset = all
   GUNIT = PCT
   rotate = landscape
   gsfmode = replace
   gsfname = GSASFILE
   device = &DRIVER
   lfactor = 1
   hsize = 8 in
   horigin = 0 in
   vsize = 6 in
   vorigin = 6 in
   ftext = "&FONTNAME"
   htext = 10pt
   ftitle = "&FONTNAME"
   htitle = 10pt
;
```

```
SYMBOL1 VALUE = None;
axis1 major = none minor = none order = (0 to 51 by 1)
  label = (h = 2.5 font = "&FONTNAME" "Subjects Sorted by IOP
Reduction");
axis2 order = (-10 to 20 by 1) minor = none label = (a = 90
r = 0 h = 2.5
  font = "&fontname" "IOP Reduction from Baseline (mm Hg)");
title1 "Figure 12.1 Waterfall Plot: IOP Reduction from
Baseline";
footnote1 "&pgmpth.";
FILENAME GSASFILE "&OUTLOC.\Figure 12.1.&EXT.";
proc gplot data = iop_reduction;
  plot reduction*subj_odr/anno = anno haxis = axis1 vaxis =
axis2 vref = 0;
  format subj_odr subjdf.;
run;
quit;

** Reproduce the figure in SGPLOT;
ods listing gpath = "&outloc.";
ods graphics/reset = all width = 8in height = 6in noborder
OUTPUTFMT = ps imagename = "FigSG_12_1";
title1 "Figure 12.1 Waterfall Plot: IOP Reduction from
Baseline";
footnote1 "&pgmpth.";
proc sgplot data = iop_reduction;
  needle x = subj_odr y = reduction1/lineattrs = (thickness = 6
PATTERN = solid COLOR = red);
  needle x = subj_odr y = reduction2/lineattrs = (thickness = 6
PATTERN = DOT COLOR = blue);
  format subj_odr subjdf.;
  xaxis display = (nolabel);
  yaxis label = 'IOP Reduction from Baseline (mm Hg)' VALUES =
(-10 to 20 by 1);
  label reduction1 = "Drug A (N = &N1.)" reduction2 = "Drug B
(N = &N2.)";
  keylegend/location = inside position = bottom;
run;
```

12.6.2 Histograms

```
*****************************************************************;
* Program Name: Chapter 12 Other Types of Plots: Histograms.sas *;
* Descriptions: Producing the following sample figures in    *;
*   chapter 12                                                *;
* - Figure 12.2 Histogram with density plots                 *;
* - Figure 12.3 Flow Charts for subject dispositions         *;
*****************************************************************;
```

```
options mprint symbolgen nodate nonumber validvarname = v7
orientation = landscape;
%let pgmname = Chapter 12 Other Types of Plots - Histograms.
sas;
%let pgmloc = C:\SASBook\SAS Programs;
%let outloc = C:\SASBook\Sample Figures\Chapter 12;
%let pgmpth = &pgmloc.\&pgmname. &sysdate9. &systime. SAS
V&sysver.;
%let seed = 12;

** Generate the age data for 200 subjects: 18 to 80 years old;
data age;
  do i = 1 to 200;
    subjid = 1000+ i;
    shuffle1 = ranuni (&SEED.);
    shuffle2 = ranuni (&SEED.+i);
    age = round((18 + ranuni (subjid)*62), 1);
    if 0 < = shuffle1 < = 0.5 then trtgrp = 1;
    if 0.5 < shuffle1 < = 1 then trtgrp = 2;
    if 0 < = shuffle2 < = 0.6 then sex = 1;
    if 0.6 < shuffle2 < = 1 then sex = 2;
    output;
  end;
  drop i;
  format trtgrp trtdf. sex sexdf.;
run;

** N, Min, Max and SD;
proc means data = age noprint;
  var age;
  output out = age_sum n = num mean = mean median = med std =
std;
run;

** Save N, Min, Max and SD to macro variables;
data _null_;
  set age_sum;
  call symput ('Num', put(Num, 5.0));
  call symput ('Mn', put(Mean, 5.1));
  call symput ('Md', put(Med, 5.1));
  call symput ('SD', put(STD, 5.1));
run;

%LET FONTNAME = Times;
%LET DRIVER = PSCOLOR;%LEt EXT = PS;
goptions
  reset = all
  GUNIT = PCT
  rotate = landscape
  gsfmode = replace
```

```
    gsfname = GSASFILE
    device = &DRIVER
    lfactor = 1
    hsize = 8 in
    horigin = 0 in
    vsize = 6 in
    vorigin = 6 in
    ftext = "&FONTNAME"
    htext = 10pt
    ftitle = "&FONTNAME"
    htitle = 10pt
;

title1 "Figure 12.2 Histogram: Distribution of Age Overall";
footnote1 "&pgmpth.";
FILENAME GSASFILE "&OUTLOC.\Figure 12.2.&EXT.";
PROC UNIVARIATE DATA = age noprint;
  HISTOGRAM age/NORMAL CFILL = ltgray;
  INSET N = 'N' MEDIAN (5.1) MEAN (5.1) STD = "SD" (5.1)/
POSITION = ne;
RUN;

title1 "Figure 12.3 Histogram: Distribution of Age by Treatment
Group and Gender";
footnote1 "&pgmpth.";
FILENAME GSASFILE "&OUTLOC.\Figure 12.3.&EXT.";
PROC UNIVARIATE DATA = age noprint;
  class trtgrp sex;
  HISTOGRAM age/NORMAL CFILL = ltgray;
  INSET N = 'N' MEDIAN (5.1) MEAN (5.1) STD = "SD" (5.1)/
POSITION = ne;
RUN;

** Reproduce the figures in SGPLOT;
%LET OUTPUTFMT = ps;
ods listing gpath = "&outloc.";
ods graphics/reset = all width = 8in height = 6in noborder
OUTPUTFMT = &OUTPUTFMT. imagename = "Fig12_2";
title1 "Figure 12.2 Histogram: Distribution of Age Overall";
footnote1 "&pgmpth.";
proc sgplot data = age noautolegend;
  histogram age/transparency =.5;
  density age;
  xaxis display = (nolabel);
  inset "N:&Num., Median:&MD., MEAN:&MN., SD:&SD." /position =
TOPRIGHT;
run;
```

```
ods graphics/reset = all width = 8in height = 6in noborder
OUTPUTFMT = &OUTPUTFMT. imagename = "Fig12_3";
title1 "Figure 12.3 Histogram: Distribution of Age by Treatment
Group and Gender";
footnote1 "&pgmpth.";
proc sgpanel data = age noautolegend;
   panelby sex trtgrp/layout = lattice novarname;
   histogram age/transparency =.5;
   density age;
   rowaxis; colaxis;
run;
```

13

Bland-Altman Plots for Agreement Analyses

13.1 Introduction

Correlation analyses between two variables are commonly used with correlation coefficient and p-value provided. However, in many cases, what we want to evaluate is the agreement between the two variables, not the correlation. Agreement is different than correlation (Bland and Altman, 1986). Variables/measurements that have good agreement will have good correlation, but the opposite is not necessary true. Two variables with perfect correlation do not necessarily have good agreement. In some cases, agreement analyses makes more sense than the correlation analyses.

Bland and Altman published a classical paper on agreement between two continuous variables and proposed to use two plots, known as the Bland–Altman plots to study the agreement between two continuous variables (Bland and Altman, 1986). The application of Bland–Altman plots for agreement analyses in clinical and medical research has gained popularity. This chapter introduces the application of Bland–Altman plots for agreement analyses using the SAS programs included in the Appendix (Section 13.7).

Bland–Altman plots for agreement analysts consist of two plots—one identity plot with an identical line and scatters of the two variables, and the other a scatter plot for the difference versus the average value of the two variables with reference lines for mean and mean +/–2*SD (standard deviation) of the difference (Figures 13.1 and 13.2).

13.2 Agreement versus Correlation

Agreement is different than correlation and the following are the main differences between the two analyses (Bland and Altman 1986, 2003).

- Correlation measures the strength of a relationship between two variables, not the agreement between them.

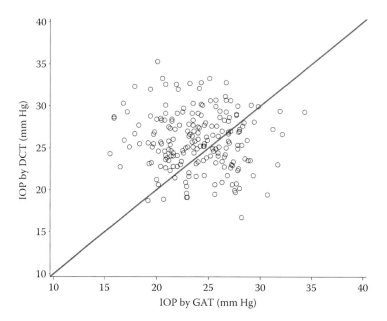

FIGURE 13.1
Agreement analyses between GAT and DCT IOP—identity plot.

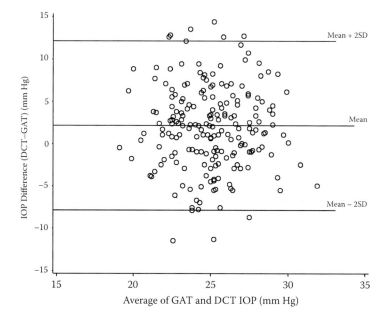

FIGURE 13.2
Agreement analyses between GAT and DCT IOP—difference verse mean value plot.

- A change in the scale of measurement does not affect the correlation, but it certainly affects the agreement.
- Correlation depends on the range of the true quantity in the sample, but agreement does not.
- The test of significance may show that the two methods are related, but it is irrelevant to the question of agreement.
- Data that seem to be in poor agreement can produce quite high correlations.

13.3 Application Example

In Chapter 6, we explored the relationship between the intraocular pressure (IOP) values measured by the Goldmann applanation tonometor (GAT) and the dynamic contour tonometer (DCT) by central corneal thickness (CCT) using thunderstorm or raindrop scatter plots. The actual agreement between GAT and DCT IOPs can be better examined using Bland–Altman plots, which are composed of an identity plot and a plot for difference versus the mean (Figures 13.1 and 13.2). In perfect agreement, the two variables would fall along the identity line in Figure 13.1 and the difference between the two variables would be close to 0 and all within the 2 reference lines of mean $+/-2*SD$. However, the plots show that the DCT and GAT IOPs are scattered quite far away from the identity line with more DCT IOP points above the line (Figure 13.1). In addition, the DCT IOP is about 2 mm Hg higher than the GAT IOP, and there are some points with differences above and below the reference lines of mean $+/-2*SD$ (Figure 13.2).

13.4 Producing the Sample Figures

13.4.1 Data Structure

We simulate 100 subjects' GAT and DCT IOP data for both eyes (OD and OS) and calculate the difference between GAT and DCT IOPs and the average of GAT and DCT IOPs. Part of the dataset is displayed in Table 13.1.

13.4.2 Notes to SAS Programs

The same Bland–Altman plots including the identity plot and the scatter plot for average versus difference are produced using both PROC GPLOT and PROC SGPLOT. The programs consist of the following main sections.

TABLE 13.1

Data Structure with Part of the Simulated
GAT and DCT IOPs

subjid	eye	iop_gat	iop_dct	diff	avg
1001	OD	21.6	26.2	4.6	23.9
1001	OS	21.9	32.8	10.9	27.4
1002	OD	26.7	29.5	2.8	28.1
1002	OS	26.3	21.1	−5.2	23.7
1003	OD	19.5	23.4	3.9	21.5
1003	OS	25.9	24.2	−1.7	25.1
1004	OD	20.8	28.5	7.7	24.7
1004	OS	22.5	24.9	2.4	23.7
1005	OD	24.8	24	−0.8	24.4
1005	OS	25.6	19.7	−5.9	22.7

13.4.2.1 Dataset Simulation and Analyses

GAT and DCT IOP values for each eye (OD and OS) of 100 subjects are simulated to have the mean values of 24 and 26 mm Hg, respectively, and have the SD of 3.5 for both GAT and DCT. The difference between the GAT and DCT IOP and the average of the two IOPs are calculated.

The mean and the SD of the difference between the GAT and DCT IOPs are calculated using *Proc Means*. The values for the mean, mean +2 SD, and mean −2 SD are saved into macro variables using *Call Symput()*. An SAS annotated dataset is produced to draw the identity line in GPLOT.

13.4.2.2 Producing the Sample Plots Using the GPLOT Procedure

- The *identity plot* is composed of a scatter plot for DCT by GAT IOP together with an identical line. *INTERPOL* = *None* option in the SYMBOL statement is used to produce a scatter for DCT by GAT IOP, and the SAS annotate facility is used to produce the identical line.

- The *difference versus the average plot* is a scatter plot for the difference versus the average of the two IOPs with reference lines for the mean, the mean −2 SD, and the mean +2 SD of the difference. The reference lines are drawn using the *VREF* = option, and the labels are specified using the *REFLABEL* = option in the AXIS2 statement.

13.4.2.3 Producing the Sample Plots Using the SGPLOT Procedure

- The *identity plot* is composed of a scatter plot for DCT by GAT IOP together with an identical line. The LINEPARM plot statement in the SGPLOT procedure is used to draw the line beginning from (0, 0) with the slope of 1.0.

- The *difference versus the average plot* is a scatter plot for the difference versus the average of the two IOPs with reference lines to indicate the positions for the mean, mean −2 SD, and mean +2 SD. The reference lines are drawn using the LINEPARM plot statements with the labels specified by the CURVELABEL option.

13.5 Summary and Discussion

The Bland–Altman plots for agreement analyses are good data visualization tools to examine the real agreement between the two continuous variables; they can be plotted using both GPLOT and SGPLOT. The identical line in the identity plot and the reference lines for the mean and the mean +/−2 SD can be drawn using the annotate facility in GPLOT or easily using the LINEPARM plot statement in SGPLOT.

13.6 References

Bland, J.M., and Altman, D.G. 1986. "Statistical Methods for Assessing Agreement between Two Methods of Clinical Measurement," *Lancet* 327: 307–310.

Bland, J.M., and Altman, D.G. 2003. "Applying the Right Statistics: Analyses of Measurement Studies," *Untrasound Obstet Gynecol* 22: 85–93.

13.7 Appendix: SAS Programs for Producing the Sample Figures

```
***************************************************************;
* Program Name: Chapter 12 Bland-Altman Plots for Agreement *;
* Analyses.sas                                               *;
* Descriptions: Producing the following sample figures in    *;
* chapter 12                                                 *;
* - Figure 13.1 Bland-Altman Plots: Identity Plot            *;
* - Figure 13.2 Bland-Altman Plots: Difference Verse Mean    *;
***************************************************************;
options mprint symbolgen nodate nonumber validvarname = v7
orientation = landscape;
%let pgmname = Chapter 13. Bland-Altman Plots for Agreement
Analyses.sas;
%let pgmloc = C:\SASBook\SAS Programs;
```

```
%let outloc = C:\SASBook\Sample Figures\Chapter 13;
%let pgmpth = &pgmloc.\&pgmname. &sysdate9. &systime. SAS
V&sysver.;

** Set the mean IOP for GAT and DCT for simulation purposes;
%let MN_GAT = 24;
%let seed = 13;
%let MN_DCT = 26;
%let N_SUBJ = 100;

proc format;
  value eyedf
     1 = 'OD'
     2 = 'OS';
  value iopdf
     1 = 'GAT'
     2 = 'DCT';
run;

** Simulate 100 subject's IOP data in both eyes;
data iop;
  do i = 1 to &N_SUBJ.; ** 100 subjects;
    do j = 1 to 2; ** 2 eyes;
       subjid = 1000 + i;
          eye = j;
          iop_gat = round((RANNOR(&seed. + i + j)* 3.5 +
&MN_GAT.),.1);
          iop_dct = round((RANNOR(&seed. + i + j)* 3.5 +
&MN_DCT.),.1);
          output;
      end;
   end;
   format eye eyedf.;
   drop i j;
run;

** Difference and average of the two types of IOPs;
data iop_diff;
  set iop;
  diff = iop_dct - iop_gat;
  avg = round(mean (iop_dct, iop_gat),.1); ** average of the 2
IOPs;
run;

** Calculate the statistics for mean and SD of the difference;
proc means data = iop_diff noprint;
   var diff;
   output out = diff_stat mean = mn_diff std = std_diff;
run;
```

```
data diff_stat;
   set diff_stat;
   drop _TYPE_ _FREQ_;
   ubound = mn_diff + 2*std_diff;
   lbound = mn_diff - 2*std_diff;
   length mn 5;
   mn = round(mn_diff, 0.01);
   call symput ('mn', mn);
   call symput ('low', lbound);
   call symput ('hih', ubound);
run;

data anno;
   function = 'move'; xsys = '1'; ysys = '1'; x = 0; y = 0;
output;
   function = 'draw'; xsys = '1'; ysys = '1'; color = 'red';
x = 100; y = 100; size = 2;
   output;
run;

%LET FONTNAME = Times;
%LET DRIVER = PSCOLOR;%LEt EXT = PS;
goptions
   reset      = all
   GUNIT      = PCT
   rotate     = landscape
   gsfmode    = replace
   gsfname    = GSASFILE
   device     = &DRIVER
   lfactor    = 1
   hsize      = 8 in
   horigin    = 0 in
   vsize      = 6.5 in
   vorigin    = 0 in
   ftext      = "&FONTNAME"
   htext      = 10pt
   ftitle     = "&FONTNAME"
   htitle     = 10pt
;

title1 "Figure 13.1";
title2 "Agreement Between GAT and DCT IOP";
title3 "Identity Plot";
footnote1 "&pgmpth.";

SYMBOL H = 4 C = BLACK     CO = BLACK I = NONE font = 'albany
amt/unicode' VALUE = '25cb'x;
FILENAME GSASFILE "&OUTLOC./Fig 13.1.&EXT";
```

```
AXIS1 MINOR = NONE ORDER = (10 TO 40 BY 5) OFFSET = (1,1)
   LABEL = (FONT = "&FONTNAME." A = 90 "IOP by DCT (mm Hg)"
H = 2.5) VALUE = (H = 2);
AXIS2 MINOR = NONE ORDER = (10 TO 40 BY 5) OFFSET = (1,1)
   LABEL = (FONT = "&FONTNAME." "IOP by GAT (mm Hg)" H = 2.5)
VALUE = (H = 2);

ods proclabel = "Agreement Between GAT and DCT IOP";
proc gplot data = iop_diff;
   plot iop_dct*iop_gat/VAXIS = AXIS1 HAXIS = AXIS2 anno = anno
      noframe nolegend des = "- Identity Plot";
run;

title1 "Figure 13.2";
title2 "Agreement Between GAT and DCT IOP";
title3 "Bland-Altman Plot: Difference Verse Mean";
footnote1 "&pgmpth.";

FILENAME GSASFILE "&OUTLOC/Fig 13.2.&EXT";
AXIS1 MINOR = NONE ORDER = (-15 TO 15 BY 5) OFFSET = (1,1)
VALUE = (H = 2)
   LABEL = (FONT = "&FONTNAME." A = 90 "IOP Difference (DCT
- GAT) (mm Hg)" H = 5.0)
   reflabel = (position = top c = blue font = "&FONTNAME."
h = 2 j = r "Mean" "Mean - 2SD" "Mean + 2SD");
AXIS2 MINOR = NONE ORDER = (15 TO 35 BY 5) OFFSET = (1,1)
VALUE = (H = 2)
   LABEL = (FONT = "&FONTNAME." "Average of GAT and DCT IOP
(mm Hg)" H = 5.0);
plot diff*avg/VAXIS = AXIS1 HAXIS = AXIS2 noframe nolegend
   vref = &mn. vref = &low. vref = &hih. des = "- Difference
Verse Mean Plot";
run;
quit;

** Reproduce using SGPLOT;
%LET OUTPUTFMT = PS;
ods listing gpath = "&outloc.";
ods graphics/reset = all width = 8in height = 6in noborder
OUTPUTFMT = &OUTPUTFMT. imagename = "Fig_13_1";

title1 "Figure 13.1";
title2 "Agreement Between GAT and DCT IOP";
title3 "Identity Plot";
footnote1 "&pgmpth.";

proc sgplot data = iop_diff noautolegend;
   scatter x = iop_gat y = iop_dct/markerattrs = (symbol = circle
color = black size = 12);
```

```
  lineparm x = 0 y = 0 slope = 1.0 /lineattrs = (color = red
pattern = solid thickness = 2);
  xaxis values = (10 to 40 by 5) label = "IOP by GAT (mm Hg)";
  yaxis values = (10 to 40 by 5) label = "IOP by DCT (mm Hg)";
run;

ods graphics/reset = all width = 8in height = 6in noborder
OUTPUTFMT = &OUTPUTFMT. imagename = "Fig_13_2";
title1 "Figure 13.2";
title2 "Agreement Between GAT and DCT IOP";
title3 "Bland-Altman Plot: Difference Verse Mean";
footnote1 "&pgmpth.";
proc sgplot data = iop_diff noautolegend;
  scatter x = avg y = diff/markerattrs = (symbol = circle
color = black size = 12);
  xaxis values = (15 to 35 by 5) label = "Average of GAT and
DCT IOP (mm Hg)";
  yaxis values = (-15 to 15 by 5) label = "IOP Difference
(DCT - GAT) (mm Hg)";
  lineparm x = 0 y = &mn. slope = 0 /lineattrs = (color =
black pattern = solid)
     curvelabel = "Mean" curvelabelpos = max curvelabelattrs =
(color = blue);
  lineparm x = 0 y = &low. slope = 0 /lineattrs = (color =
black pattern = solid)
     curvelabel = "Mean - 2SD" curvelabelpos = max
curvelabelattrs = (color = blue);
  lineparm x = 0 y = &hih. slope = 0 /lineattrs = (color =
black pattern = solid)
     curvelabel = "Mean + 2SD" curvelabelpos = max
curvelabelattrs = (color = blue);
run;
quit;
```

14

SAS ODS Graphic Designer

14.1 Introduction

This chapter illustrates how to use SAS ODS Graphics Designer to produce high-quality graphs. SAS ODS Graphics is a graphical user interface (GUI)-based interactive tool for creating custom graphs quickly without any programming (SAS Institute Inc., 2011). It was an experimental feature in SAS 9.2, is part of the formal release for SAS 9.3, and is part of the base SAS. The designer is a good tool for quickly producing some exploratory graphics before formal analysis (Matange, 2009, 2012). It can also help us to learn SAS graphic programming in the Graph Template Language (GTL) because the designer creates graphs that are based on the GTL, and the SAS codes in GTL can be viewed automatically.

14.2 SAS ODS Graphic Designer

14.2.1 The Designer GUI

The graphic designer is started by invoking the SAS built-in macro *%sgdesign* or *%sgdesign()* in the program editor window. After running the macro, a separate ODS Graphics Designer GUI window will pop up (Figure 14.1). The design window is composed of the following panels and bars.

- The Menu bar: located along the top, with File, Edit, View, Insert, Format, Tools, and Help bars.
- The Toolbar: common items similar to the ones found in Microsoft Word.
- The Graph Gallery: commonly used graphs organized in tabs.
- Plots and Insets: on the left side, initially inactive and activated after a blank graph is created.

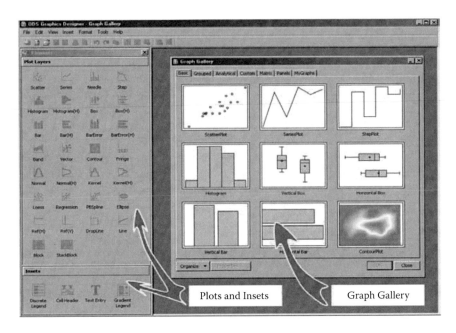

FIGURE 14.1
The layout of the ODS Graphics Designer Interface. (Figure 2, Matange, Sanjay. 2009. "ODS Graphics Designer: An Interactive Tool for Creating Batchable Graphs," *Proceedings of the SAS Global Forum 2009 Conference,* Cary, NC: SAS Institute Inc., https://support.sas.com/resources/ papers/proceedings09/331-2009.pdf.) Courtesy: SAS Institute, Inc. Cary, NC, USA.

14.2.2 The Graph Gallery

The graph gallery contains a tabbed set of commonly used graphs:

- Basic: Common graphs
- Grouped: Graphs showing grouped data
- Analytical: Graphs commonly used for analysis of data
- Custom: A set of graphs showing the possible ways to combine the plots
- Matrix: A set of scatter plot matrix graphs
- Panels: A set of classification panel graphs

14.3 Using the ODS Graphic Designer to Reproduce Some Figures in the Previous Chapters

Let's use SAS ODS Graphics Designer to reproduce a line-up jittered scatter plot (Figure 14.2) and a classification box plot (Figure 14.3), as described in

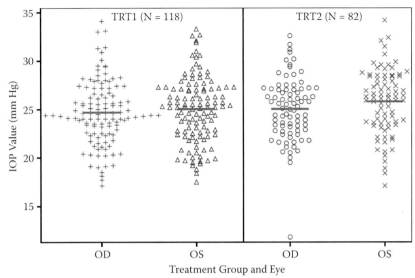

FIGURE 14.2
Line-up jittered scatter plot reproduced in ODS Graphic Designer.

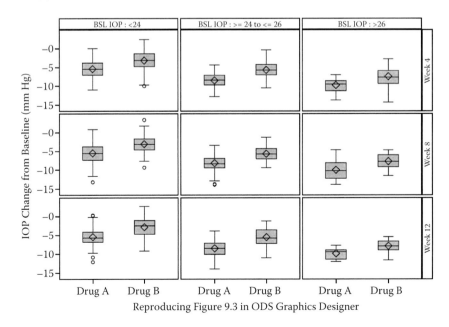

FIGURE 14.3
Classification box plot for treatment effect by baseline IOP and visit reproduced in ODS Graphic Designer.

Chapters 5 and 9, without using the codes. Please note that datasets must be produced and manipulated before figures can be generated in the designer using an interactive drag-and-drop process.

14.3.1 Line-Up Jittered Scatter Plots

Figure 14.2 (line-up jittered scatter plot for IOP by treatment and eye) described in Chapter 5, can be reproduced in ODS Graphics Designer by following these steps.

1. Run part of the SAS programs in Chapter 5 to prepare the datasets; the codes are included in the Appendix (Section 14.6.1).
2. Manipulate and prepare a dataset that is ready to be used in ODS Graphics Designer: A dataset *line_mn* is prepared to contain the x and y coordinates for the four mean values of the two treatment groups and two eyes. It is then set with the dataset *iop_jittered* to prepare a final dataset called *iop_jscatter* that is ready for the designer.

```
data line_mn;
  set iop_stat;
  if index = 1 then do;
    x1 = index -.2; y1 = mean; output;
    x1 = index +.2; y1 = mean; output;
  end;
  if index = 2 then do;
    x2 = index -.2; y2 = mean; output;
    x2 = index +.2; y2 = mean; output;
  end;
  if index = 3 then do;
    x3 = index -.2; y3 = mean; output;
    x3 = index +.2; y3 = mean; output;
  end;
  if index = 4 then do;
    x4 = index -.2; y4 = mean; output;
    x4 = index +.2; y4 = mean; output;
  end;
  keep x: y:;
run;

data iop_jscatter;
  set iop_jittered line_mn;
    format index indexdf.;
run;
```

3. Select **Scatter Plot** from the Graph Galley, and then select **OK**.

4. Choose the data and plot variables from the Assign Data dialogue window:

   a. Library: *WORK*

   b. Data Set: *IOP_JITTERED*

   c. X: *XVAR_J*

   d. Y: *IOP*

   e. Group: *INDEX*

5. Update the x- and y-axis labels: the variable names are used in default; we can double-click to update them directly.

   a. x-axis label: Treatment Group and Eye

   b. y-axis label: *IOP Value (mm Hg)*

6. Add short red lines to denote the mean IOP value for all treatment and eye combinations.

   a. Right-click in the graph area and choose **Add an Element**: a new Elements window pops up.

   b. Choose **Series**; the Assign Data dialogue box appears. The Library and Data Set fields are preselected to be the same as in the original graph.

   c. Choose variables: X: *x1*, Y: *y1*. A short line will be overlaid to denote the mean value for the OD eye in TRT1 group.

   d. Enter this name for series plot: *series1*.

   e. Repeat steps a through d to overlay short series lines to denote mean values for other eyes and treatments (*series2, series3,* and *series4*).

7. Set the plot properties: Right-click the graph area and select **Plot Properties**.

   a. Scatter plot: Select **Scatter** from the plot type. The color, symbol, size, and weight can be reset from the drop-down menus.

   b. Series: Select **series1** from the plot type and reset the color (red), pattern (solid line), and thickness (3) from the drop-down menus.

   c. Repeat steps a and b to set the same properties for *series2* to *series4*.

8. Add a vertical reference to separate scatters in the two treatment groups.

   a. Drag the *Ref(V)* icon from the Plot Layers panel to the graph (can also be done by right-clicking the graph and choosing Add an Element).

   b. Enter **2.5** for the x value in the pop-up Assign Data - Ref(V) window.

9. Add text for TRT1 (N = 112) and TRT2 (N = 82).

   a. Drag the *Text Entry* icon from the Plot Layers panel to the graph (can also be done by right-clicking the graph and choosing Text Entry).

   b. Type in **TRT1 (N = 112)** and **TRT2 (N = 82)** and position them in
       the desired locations.

10. Type in titles and footnotes:

   a. Title: Figure 14.2 Line-up Jittered Scatter Plot for IOP by Treatment
       and Eye.

   b. Footnotes: Footnote1 "Note: Red line denotes the mean value
       within each treatment and eye" and footnote2 "Reproducing
       Figure 5.2 in ODS Graphics Designer."

11. Save the graph.

   a. Click **File** > **Save As**: The Save As dialogue box appears.

   b. The file type is *.SGD (DGD Files)* by default. We can assign a
       name, choose a location, and click **OK** to save the output as SGD.

   c. We can also select other file types (JPG, BMP, PNG, PDF, PS, and
       EMF). For JPG and PNG, we can select a resolution in DPI. PDF,
       PS, and EMF are vector format with high quality.

12. View SAS codes: Click **View** > **Code**: The codes can be selected
    and copied.

14.3.2 Classification Box Plots

Figure 14.3, a classification box plot with treatment effect by baseline IOP and
visits (discussed in Chapter 9), can be reproduced in ODS Graphics Designer
by following these steps:

1. Run part of the SAS programs in Chapter 9 to prepare the dataset
    mndiur_chg: Codes are included in the Appendix (Section 14.6.3).

2. Create a new blank graph. Drag the box plot from the Plot Layers
    panel and drop it into the graph work area.

3. Choose the data and plot variables from the Assign Data dialog window:

   a. Library: *WORK*

   b. Data Set: MNDIUR_CHG

   c. X: TRTNUM

   d. Y: *CHG*

4. Click **Panel Variables** within the Assign Data window.

   a. Choose either Data Lattice or Data Panel. Let's select **Data Lattice**.

   b. Column: *BSL_C*

   c. Row: *VISIT*

5. Click **OK** and a box plot classified by baseline IOP category (column)
    and visit (row) will be produced.

6. Update the x- and y-axis labels: The variable names are used by default and can be updated by being double-clicked.

   a. x-axis label: No need for a label, so we can delete the default one.

   b. y-axis label: IOP Change from Baseline (mm Hg)

7. Set Axis properties: Move the cursor to the x-axis area. Right-click and select **Axis Properties**.

   a. Data Range: Select **Union** from the drop-down menu and the empty space for treatment will disappear.

   b. Label, value, grid and tick, and so on can be updated in this window if needed.

8. Insert titles and footnotes: Click **Insert** in the menu bar and select **Title** and **Footnote**.

   a. Title: Figure 14.3 Treatment Effect by Baseline IOP and Visit

   b. Footnote: Reproducing Figure 9.3 in ODS Graphics Designer

9. Save the graph.

   a. Click **File** > **Save As**: the Save As dialog box appears.

   b. By default, the file type is *SGD (DGD Files)* by default. Assign a name **Producing Box Plot in Classification Panel in ODS Graphic Designer.sgd**, choose a location, and click **OK** to save the output as SGD.

   c. We can also select other file types (JPG, BMP, PNG, PDF, PS, and EMF). For JPG and PNG, we can select a resolution in DPI. PDF, PS, and EMF are vector format and have high quality.

10. View SAS codes: Click **View** > **Code**: The codes can be selected and copied, and are included in the Appendix (Section 14.6.4).

14.4 Summary and Discussion

Some of the advantages of using ODS Graphic Designer to produce figures include the fact that no programming and no annotate dataset are needed to produce advanced custom graphics. The Insets panel provides an easy way to position text, legends, and cell headers in a graph.

The line-up jittered scatter plot reproduced in the designer (Figure 14.2) is a little different than the original one (Figure 5.2) in that the treatment and eye index are displayed using the group option (*Group = Index*) with different symbols in the designer. This can be changed in the cell properties to have the same symbol if you prefer.

The box plot that is reproduced in ODS Graphics Designer has two classification panels (by baseline IOP category and visit, Figure 14.3) versus one classification panel in the original figure (by visit, Figure 9.3) with the baseline IOP category that is displayed in the category in SGPANEL. Both use the lattice layout. The codes for generating the paneled plots in the designer cannot be displayed because there is a comment "The graph contains paneled computed plot(s). This is not supported by PROC SGRENDER. You need to run PROC SGDESIGN to generate the graph" in the view code window.

The saved SGD figures can be opened later for review and edits, but before doing that, you must first run the SAS programs that produce the dataset first.

14.5 References

Matange, Sanjay. 2009. "ODS Graphics Designer: An Interactive Tool for Creating Batchable Graphs." In *Proceedings of the SAS Global Forum 2009 Conference*. Cary, NC: SAS Institute Inc., https://support.sas.com/resources/papers/proceedings09/331-2009.pdf.

Matange, Sanjay. 2012. "Quick Results with SAS® ODS Graphics Designer." In *Proceedings of the SAS Global Forum 2012 Conference*. Cary, NC: SAS Institute Inc., http://support.sas.com/resources/papers/proceedings12/153-2012.pdf.

SAS Institute Inc. 2011. *SAS® 9.3 ODS Graphics Designer: User's Guide*. Cary, NC: SAS Institute Inc.

14.6 Appendix: SAS Programs for Producing the Datasets and Generated by the Designer

14.6.1 The Jittered Scatter Plot: Codes Producing the Dataset

```
** setting-up macros for data simulation;
%let seed = 05;%let subjnum = 200;%let mean = 25;%let sd 3.5;
proc format;
  value trtdf
     1 = 'TRT1'
     2 = 'TRT2'
     OTHER = ' ';
  value eyedf
     1 = 'OD'
     2 = 'OS'
     OTHER = ' ';
```

```
  value indexdf
    1, 3 = 'OD'
    2, 4 = 'OS'
    OTHER = ' ';
run;

** Simulate subject's IOP data in both eyes;
data iop;
  do i = 1 to &subjnum.; ** num. of subjects;
    if ranuni (&seed.) < = 0.5 then trt = 1;
      else trt = 2;
    do j = 1 to 2; ** 2 eyes;
      subjid = 1000 + i;
        eye = j;
        iop = round((RANNOR(&seed. + i + j)* &sd. + &mean.),.1);
        output;
    end;
  end;
  format eye eyedf. trt trtdf.;
  drop i j;
run;

** Create an index variable to combine the treatment group and
eye;
data iop;
  set iop;
  if trt = 1 and eye = 1 then index = 1;
  if trt = 1 and eye = 2 then index = 2;
  if trt = 2 and eye = 1 then index = 3;
  if trt = 2 and eye = 2 then index = 4;
run;

** Number of subjects by treatment;
proc sort data = iop out = subj_trt nodupkey;
  by trt subjid;
run;
proc freq data = subj_trt noprint;
  table trt/out = trt_freq (drop = percent);
run;

** save the subject number at each treatment group to macro
variables;
data _null_;
  set trt_freq;
  if trt = 1 then call symput ('num_trt1', put(count, 3.0));
  if trt = 2 then call symput ('num_trt2', put(count, 3.0));
run;
```

```
%macro LineUp_Jitter (idn =, xvar =, yvar =, lvl =, jitter =,
odn =);
** Get the Min and Max values and save them into macro
variables;
PROC MEANS DATA = &idn. MIN MAX maxdec = 2 noprint;
  VAR &yvar.;
  OUTPUT OUT = tmp_STAT min = min max = max;
RUN;
data _null_;
  set tmp_stat;
  call symput('Min', min);
  call symput('Max', max);
run;

proc sort data = &idn. out = d_tmp;
  by &yvar.;
run;

data add_tmp;
  set d_tmp;
  do i = floor(&min.) to ceil(&max.) by &lvl.;
    if i < = &yvar. < i+1 then yvar_int = i;
  end;
run;

** get the number of data points at each interval by x-axis
value;
proc freq data = add_tmp noprint;
  tables &xvar.*yvar_int/out = freqdata;
run;

data jitter_sum;
  retain pos_neg 1;
  set freqdata;
  pos_neg = 1;

  if count = 1 then do;
    xvar_j = &xvar.;
    output;
  end;

  if count>1 then do;
    if mod(count,2) > 0 then do; ** odd count number;
      xvar_j = &xvar.; pos = &xvar.; neg = &xvar.; output;
      do i = 1 to count-1 by 1;
        if pos_neg = 1 then do;
          xvar_j = neg-&jitter;
            neg = xvar_j;
        end;
```

```
         else do;
            xvar_j = pos+&jitter;
               pos = xvar_j;
         end;
         output;
         pos_neg = -1*pos_neg;
      end;
   end;

   if mod(count,2) = 0 then do; ** even count number;
      pos = &xvar.; neg = &xvar.;
      do i = 1 to count by 1;
         if pos_neg = 1 then do;
            xvar_j = neg-&jitter/2;
               neg = xvar_j-&jitter/2;
         end;
         else do;
            xvar_j = pos+&jitter/2;
               pos = xvar_j+&jitter/2;
         end;
         output;
         pos_neg = -1*pos_neg;
      end;
   end;
 end;
run;

proc sort data = add_tmp;
   by yvar_int &xvar.;
run;

proc sort data = jitter_sum;
   by yvar_int &xvar.;
run;

** Output dataset: Jittered dataset for plotting;
data &odn.;
   merge jitter_sum (in = a) add_tmp;
   by yvar_int &xvar.;
   if a;
   keep &xvar. &yvar. xvar_j yvar_int;
run;
%mend LineUp_Jitter;

** Producing line-up jittered datasets;
** Jitter Level on X-axis = 0.08, Line-up Interval on y-axis =
0.5 mm Hg;
%LineUp_Jitter (idn = IOP, yvar = iop, xvar = index, lvl = 0.5,
jitter = 0.08, odn = IOP_Jittered);
```

```
* Summary Statistics by index (i.e., each eye at each treatment);
PROC MEANS DATA = IOP NWAY MEAN STD STDERR MIN MAX MEDIAN
maxdec = 2 noprint;
  CLASS index;
  VAR iop;
  OUTPUT OUT = iop_STAT MEAN = Mean;
RUN;

* Prepare dataset ready for ODS designer;
data line_mn;
  set iop_stat;
  if index = 1 then do;
    x1 = index -.2; y1 = mean; output;
    x1 = index +.2; y1 = mean; output;
  end;
  if index = 2 then do;
    x2 = index -.2; y2 = mean; output;
    x2 = index +.2; y2 = mean; output;
  end;
  if index = 3 then do;
    x3 = index -.2; y3 = mean; output;
    x3 = index +.2; y3 = mean; output;
  end;
  if index = 4 then do;
    x4 = index -.2; y4 = mean; output;
    x4 = index +.2; y4 = mean; output;
  end;
  keep x: y:;
run;

data iop_jscatter;
  set iop_jittered line_mn;
  format xvar_j indexdf.;
run;
```

14.6.2 The Jittered Scatter Plot: GTL Codes Generated by the Designer

```
proc template;
define statgraph sgdesign;
dynamic _XVAR_J _IOP _INDEX _X1 _Y1 _X2 _Y2 _X33 _Y3 _X42 _Y42;
begingraph;
  entrytitle halign = center 'Figure 14.2 Line-up Jittered
Scatter Plot for IOP by Treatment and Eye';
  entryfootnote halign = center 'Nore: Red lin denotes the
mean value within each treatment and eye';
  entryfootnote halign = center 'Reproducing Figure 5.2 in ODS
Graphics Designer';
  layout lattice/rowdatarange = data columndatarange = data
rowgutter = 10 columngutter = 10;
```

```
      layout overlay/xaxisopts = (display = (LINE TICKVALUES
LABEL TICKS) griddisplay = off label = ('Treatment Group and
Eye')) yaxisopts = (label = ('IOP Value (mm Hg)'));
         scatterplot x = _XVAR_J y = _IOP/group = _INDEX name =
'scatter';
         seriesplot x = _X1 y = _Y1/name = 'series' connectorder
= xaxis lineattrs = (color = CXFF0000 pattern = SOLID
thickness = 3);
         seriesplot x = _X2 y = _Y2/name = 'series2'
groupdisplay = Cluster connectorder = xaxis lineattrs = (color
= CXFF0000 thickness = 3);
         seriesplot x = _X33 y = _Y3/name = 'series3'
connectorder = xaxis lineattrs = (color = CXFF0000 pattern =
SOLID thickness = 3);
         seriesplot x = _X42 y = _Y42/name = 'series4' connectorder
= xaxis lineattrs = (color = CXFF0000 thickness = 3);
         referenceline x = 2.5/name = 'vref' xaxis = X;
         referenceline x = 2.5/name = 'vref2' xaxis = X;
         entry halign = left 'TRT1 (N = 118)'/valign = top;
         entry halign = center 'TRT2 (N = 82)'/valign = top;
      endlayout;
   endlayout;
endgraph;
end;
run;

proc sgrender data = WORK.IOP_JSCATTER template = sgdesign;
dynamic _XVAR_J = "'XVAR_J'n" _IOP = "IOP" _INDEX = "INDEX"
_X1 = "X1" _Y1 = "Y1" _X2 = "X2" _Y2 = "Y2" _X33 = "X3"
_Y3 = "Y3" _X42 = "X4" _Y42 = "Y4";
run;
```

14.6.3 The Classification Box Plot: Codes for Producing the Dataset

```
** Set-up macro variables for data simulation;
%LET SD = 3.5;
%let sitenum = 10;
%let seed = 08;
%let subjnum = 500;

proc format;
   value trtdf
      1, 4, 7 = 'Drug A'
      2, 5, 8 = 'Drug B'
      OTHER = ' ';
   value visdf
      1 = 'Baseline'
      2 = 'Week 4'
      3 = 'Week 8'
      4 = 'Week 12';
```

```
    value hrdf
       1 = 'Hour 0'
       2 = 'Hour 2'
       3 = 'Hour 8';
    value bsldf
       1 = 'BSL IOP: < 24'
       2 = 'BSL IOP: > = 24 to < = 26'
       3 = 'BSL IOP: > 26'
       OTHER = ' ';
run;

** Generate the required number of subjects and randomly
assign to 2 treatment groups;
data subj;
   do i = 1 to &subjnum.;
      subjid = 1000+ i;
      if ranuni (&seed.) < 0.5 then trtnum = 1;
         else trtnum = 2;
      output;
   end;
   drop i;
run;

** Set up the IOP Values based on the trt assignment and
visits/timepoints;
data iop;
   set subj;
   do i = 1 to 4; ** 4 visits;
      do j = 1 to 3; ** 3 timepoints/visit;
         visit = i;
         hour = j;
         if i = 1 then do; ** Baseline;
            if j = 1 then iop = round((RANNOR(&seed.)* &SD.
+ 25),.1); ** Hour 0;
            if j = 2 then iop = round((RANNOR(&seed.)* &SD.
+ 23),.1); ** Hour 2;
            if j = 3 then iop = round((RANNOR(&seed.)* &SD.
+ 22),.1); ** Hour 8;
         end;
         else if i > 1 and trtnum = 1 then do; ** Post-baseline:
drug A;
            if j = 1 then iop = round((RANNOR(&seed.)* &SD.
+ 17.5),.1); ** Hour 0;
            if j = 2 then iop = round((RANNOR(&seed.)* &SD.
+ 16.5),.1); ** Hour 2;
            if j = 3 then iop = round((RANNOR(&seed.)* &SD.
+ 16.2),.1); ** Hour 8;
      end;
```

```
      else if i > 1 and trtnum = 2 then do; ** Post-baseline:
drug B with 2.5 mmHg higher at each point;
        if j = 1 then iop = round((RANNOR(&seed.)* &SD.
+ 20),.1); ** Hour 0;
        if j = 2 then iop = round((RANNOR(&seed.)* &SD.
+ 19),.1); ** Hour 2;
        if j = 3 then iop = round((RANNOR(&seed.)* &SD.
+ 18.7),.1); ** Hour 8;
    end;
    output;
      end;
    end;
  drop i j;
  format visit visdf. hour hrdf. trtnum trtdf.;
run;

proc sort data = iop;
  by subjid visit hour;
run;

** Mean diurnal IOP at each visit: mean of hours 0, 2 and 8
average eye IOP at baseline;
proc transpose data = iop out = iop_t;
  by subjid trtnum visit;
  var iop;
  id hour;
run;

data iop_MnDiur;
  set iop_t;
  iop_MnDiur = round (mean (Hour_0, Hour_2, Hour_8),.01);
  drop _NAME_ HOUR_:;
run;

** baseline mean ddiurnal IOP and category;
data mndiur_bsl;
  set iop_MnDiur;
  where visit = 1;
  if iop_MnDiur < 24 then bsl_c = 1;
  if 24 < = iop_MnDiur < = 26 then bsl_c = 2;
  if iop_MnDiur > 26 then bsl_c = 3;
  rename iop_MnDiur = mndiur_bsl;
  format bsl_c bsldf.;
  drop visit;
run;

** change from baseline;
data mndiur_chg;
  merge iop_MnDiur (where = (visit > 1)) mndiur_bsl;
  by subjid;
```

```
    chg = iop_MnDiur - mndiur_bsl;
run;
proc sort data = mndiur_chg;
  by visit bsl_c trtnum;
run;
```

14.6.4 The Classification Box Plot: GTL Codes Generated by the Designer

```
*********************************************************;
* GTL Codes Generated by the Designer                  *;
*********************************************************;
proc template;
define statgraph Graph;
dynamic _TRTNUM _CHG _VISIT2 _BSL_C2;
dynamic _panelnumber_;
begingraph;
  entrytitle halign = center 'Figure 14.3 Treatment Efefct by
Basleine IOP and Visit';
  entryfootnote halign = center 'Reproducing Figure 9.3 in ODS
Graphics Designer';
  layout datalattice rowvar = _VISIT2 columnvar = _BSL_C2/
cellwidthmin = 1 cellheightmin = 1 rowgutter = 3 columngutter
= 3 rowdatarange = unionall row2datarange = unionall
columndatarange = union column2datarange = unionall
headerlabeldisplay = value columnaxisopts = (display = (TICKS
TICKVALUES)) rowaxisopts = (label = ('IOP Change from Baseline
(mm Hg)'));
     layout prototype/walldisplay = (OUTLINE FILL);
        boxplot x = _TRTNUM y = _CHG/name = 'box' capshape =
Serif boxwidth = 0.4 groupdisplay = Cluster clusterwidth =
0.7;
     endlayout;
  endlayout;
endgraph;
end;
run;

/* The graph contains paneled computed plot(s). This is not
supported by PROC SGRENDER. You need to run PROC SGDESIGN to
generate the graph. */
```

15

Producing Figures in the SAS Enterprise Guide Environment

15.1 Introduction

SAS Enterprise Guide (EG) software is a powerful Windows application that provides a guided mechanism to exploit the power of SAS and publish dynamic results throughout an organization (SAS Institute, 2010). SAS EG uses SAS software behind the scenes by either local access to SAS on your computer or remote access to SAS on another computer through an SAS server. Since SAS EG uses SAS as an engine, all the SAS programs in the book can be run in the SAS EG environment to produce high-quality figures. This chapter discusses how to create an SAS EG project to include all the SAS programs to produce the high-quality sample figures. Some pros and cons in running the SAS programs in the EG environment are discussed.

SAS EG interface is composed of three area windows: the project tree, the server list window, and the workspace area (Figure 15.1). The workspace area displays the SAS programs, data, logs, results, and process flows. Unlike standalone SAS installed in personal computers (PC SAS), which has a separate GRAPH window to display figures produced in most SAS/GRAPH procedures and a Result View to display figures in ODS Graphics procedures, SAS EG does not provide a separate window to display the graphics results. However, we can display the graphics in different result formats, like *Results-HTML, Results-PDF,* and *Results-RTF.*

15.2 Create, Run, and Save an EG Project

To create the SAS EG project to produce all the sample figures in the book, follow these steps:

1. Click **File**, **New**, and then **Project**, you will see an EG workspace as shown in Figure 15.1.

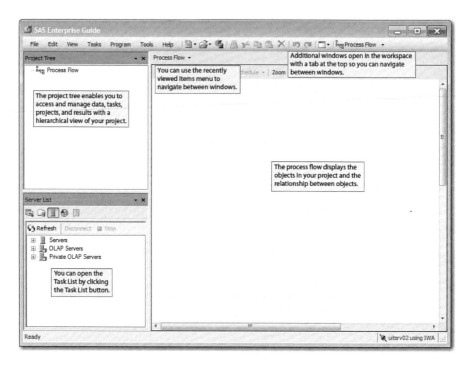

FIGURE 15.1

SAS EG window. ("Getting Started with SAS Enterprise Guide: Learning the Basics," SAS Institute Inc., http://support.sas.com/documentation/onlinedoc/guide/tut51/en/m1_1.htm.) Courtesy: SAS Institute, Inc. Cary, NC, USA.

2. Open all the programs that were used to produce the sample figures from Chapters 3 to 13. Save the project as "Book on SAS Graphics. egp" (Figure 15.2).

3. Set up the graphics options:

   a. Click **Tools** and then **Options**; a separate Options window pops up to allow you to select the options.

   b. By default, results in SAS EG, including the graphics, are displayed in HTML format. If you want to display the figures in other formats, you can choose PDF, RTF, and SAS Report.

   c. In the Graph option under the Results, you can choose different formats: ActiveX, Java, GIF, JPEG ActiveX image, JPEG image, SAS EMF, and PNG.

   d. If the SAS Report, HTML, PDF, RTF, or Text Output Result Format is checked in Options > Results > Results General, you might see warning messages (shown below) when using the device driver (PSCOLOR or SASEMF) to save listing format (PS or EMF) figures. This is because PNG is the default device driver in SAS

FIGURE 15.2
SAS EG project to produce graphics.

EG. You can either ignore these warnings, since they do not affect the figure files (PS, EMF, etc.) saved, or uncheck any result format.

- WARNING: Unsupported device 'PSCOLOR' for HTML (EGHTML) destination. Using default device 'PNG'.
- WARNING: Unsupported device 'SASEMF' for HTML (EGHTML) destination. Using default device 'PNG'.
- WARNING: Unsupported device 'PSCOLOR' for TAGSETS. SASREPORT13(EGSR) destination. Using default device 'PNG'.

4. Run the SAS programs to produce figures:

   a. Click **Run** in the Program window.

   b. The logs are displayed in the Log window as in PC SAS.

   c. All the datasets generated are arranged and displayed in the Output Data tab. Click the tab and select the dataset from the drop-down menu to view the dataset.

   d. Figures produced will be displayed in the prechosen or the produced formats (Results - HTML, Results - PDF, Results - RTF, etc.) (Figure 15.3).

5. Save the EG Project:

   a. Click **File**, then click **Save Project As**.

   b. Choose a location to save the project, enter a file name, and click **Save**. An SAS EG Project is saved in EGP (Enterprise Guide Project) format by default.

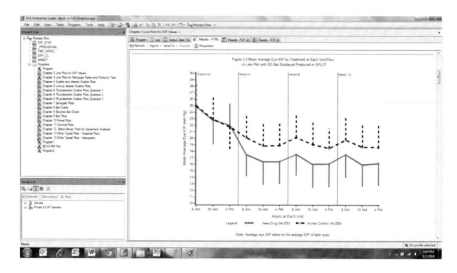

FIGURE 15.3
Figures displayed in results—HTML.

15.3 Server List

The server list panel in SAS EG contains the Libraries and Files categories.

The Work directory is located in the Libraries category: click the Work directory to see the datasets generated in the program. Move the cursor to the WORK directory, right-click, and choose **Refresh** to display all new and updated datasets.

15.4 Some Good Features and Notes for SAS EG

After opening a dataset in SAS EG, you can use the Send To icon (Figure 15.4) in the toolbox to send the dataset to Microsoft applications including Excel, Word, and so on. This is a useful feature to export a dataset to share with a third party.

The following are some noteworthy features in SAS EG:

- LOG is automatically cleared and updated every time SAS codes are run.
- A dataset can be updated even when it is open in SAS EG: when the codes producing the dataset are run, the opened dataset is automatically updated. In PC SAS, when a dataset in WORK is opened, it cannot be updated and results in an error message in the Log.

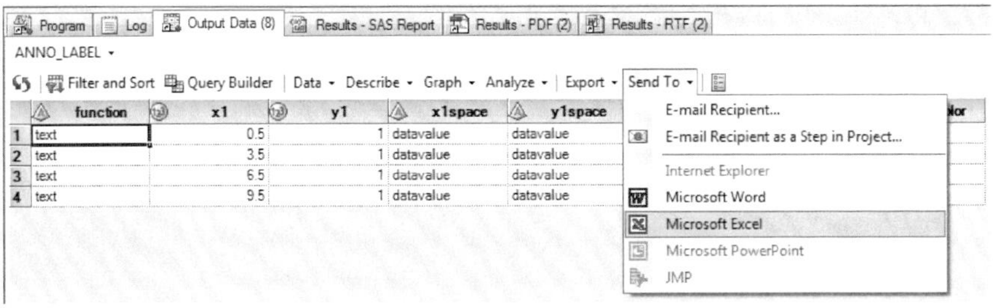

FIGURE 15.4
Send the dataset to some Microsoft applications.

- When entering data in the EG editor, SAS EG does not like data separated by tabs; you must use space instead.
- It is suggested that you use the option validvarname = v7 at the beginning of the programs, otherwise programs with *Proc Transpose* might not work as expected. By default, EG uses options validvarname = any, which allows column names with embedded spaces, special characters, and leading numerics.
- Make sure that the right tool options are chosen, otherwise some individual figures (PS or EMF) might not be produced using the filename statement in GPLOT. It is suggested that you turn on the ODS LISTING.
- As you type, some key words automatically appear for you to choose.
- You can move the cursor over the SAS keywords, for example, SGPLOT, and a window will pop up displaying contents like Syntax, Product Documentation, Sample & SAS Notes, Papers, and so on.

15.5 Producing Graphics Directly in SAS EG Using the Point-and-Click Feature

SAS EG can also produce figures directly using the point-and-click feature, without using codes after a dataset is generated and opened. This is useful for viewing and exploring some simple figures before you decide whether to do a formal analysis or graphs. Follow these steps to produce the simple line plot in Figure 15.5, similar to Figure 3.1 discussed in Chapter 3.

1. Run part of the SAS programs in Chapter 3 to prepare the dataset MN_IOP.
2. Open the MN_IOP dataset by choosing the dataset from the Output Data tab in the workspace area or by opening it from the Work directory in the Server List area.

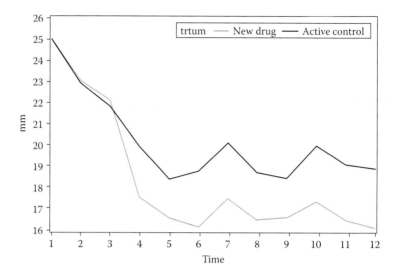

FIGURE 15.5
A simple line plot produced in EG using the point-and-click feature.

3. Click **Graph**, choose **Line Plot**, and choose **Multiple Line Plots by Group Column**.

4. Data: Assign variables for horizontal, vertical, and group in the Columns to Assign panel.

   a. Select **Time**, click the right-pointing arrow, and choose **Horizontal**. This is the time variable (visit and hours) on the x-axis.

   b. Select **mn**, click the right-pointing arrow, and choose **Vertical**. This is the response variable (mean IOP) on the y-axis.

   c. Select **trtnum**, click the right-pointing arrow, and choose **Group**. This is the group variable for the treatment.

5. Run: Click **Run** and a simple line plot by treatment group is generated (Figure 15.4).

6. View Properties: Click **Properties** in the toolbar and the properties associated with the newly generated figure will be shown. These properties cannot be edited, but the different styles can be chosen.

   a. Label: This style is the name of the results item in the project tree, e.g., HTML - Line Plot1.

   b. Created: This style is the date when the figure was generated.

   c. Last Modified: This style is the date when the figure was last modified.

   d. Modified by: This style is the person who last modified the result item.

  e. Style: By default, HTMLBlue is used. You can choose another style from the drop-down menu, such as Listing or Journal.

 7. Save/copy the figure: Right-click the graph area and choose to Copy or Save Picture AS to copy and save the figure. Note that you can only save the figure to PNG, a bitmap format that is not of high quality.

 8. Codes: You can view the codes provided in EG by clicking Code in the toolbar area.

The figures produced in this manner are for exploratory purposes only and cannot be formatted or saved in a high-quality file format. If needed, a more customized, high-quality version can be produced as described in Chapter 3 (Figure 3.1).

15.6 Summary and Discussion

In SAS EG, you can create a project that includes all SAS programs. SAS EG has some advantages compared with PC SAS in that SAS EG has an advanced program editor and the results (graphics) can be saved in different output formats (HTML, PDF, RTF, etc.) in a project for later view. SAS EG also offers a point-and-click feature to produce some simple plots.

15.7 References

SAS Institute, Inc. 2013. "Getting Started with SAS® Enterprise Guide®," http://support.sas.com/documentation/onlinedoc/guide/tut51/en/m1_1.htm.

SAS Institute, Inc. 2010. *SAS® Enterprise Guide® for Experienced SAS Programmers Course Notes*. Cary, NC: SAS Press.

Index

Printed and bound by CPI Group (UK) Ltd, Croydon, CR0 4YY

23/10/2024

01777709-0002